# WEATHERMAN

## JOHN KETTLEY

GREAT NORTHERN

Great Northern Books
PO Box 213, Ilkley, LS29 9WS
www.greatnorthernbooks.co.uk

ISBN: 978 1 905080 61 8

Design and layout: David Burrill

The publishers acknowledge the help of Yorkshire Post Newspapers [YPN] in providing photographs for this book.

CIP Data
A catalogue for this book is available from the British Library.

# CONTENTS

The never-to-be-forgotten morning of 2nd February 2009 when Londoners woke up to some 15 centimetres of snow. The city ground to a halt – and faced the reality that the weather had bucked the trend towards warmer winters.  (PA Wire)

# PROLOGUE

## Winter 2008-09

We constantly hear that "winters aren't what they used to be" and after a succession of mild winters since the start of the millennium it would be difficult to argue. On 25th September 2008 the Met Office issued their seasonal forecast for the months ahead based primarily on the influence of global ocean temperatures on European winter climate. The summary of their expectations was also very much in line with those issued by other weather forecasting organisations at the time. The banner headline read: "Trend of mild winters continues".

In a little more detail: "For the UK as a whole, winter mean temperatures are more likely to be above normal. Although a winter milder than the 1971-2000 average is favoured, temperatures are likely to be lower than those experienced last year. Rainfall is likely to be lower than observed in last year's relatively wet winter. However, this signal is not sufficient to indicate whether winter precipitation totals are more likely to be above or below the 1971-2000 average."

Vague, perhaps, but understandably the information was welcomed by Help the Aged. Dr. James Goodwin, their Head of Research, commented: "The onset of winter causes significant anxiety among many older people. This forecast will assist policy makers to adapt their strategies to ensure that the negative effects of winter weather are reduced as far as possible."

Considering the winter of 2007-2008 was the ninth mildest and the fourteenth wettest since 1914 it would come as no surprise if the 2008-2009 winter would feel somewhat colder at times and probably not as wet!

What we had in the end was totally unexpected and for some people a throwback to bygone days, although clearly not on the scale of the infamous winters of 1947 and 1963. Even so the heavy snowfalls in southern England were widely agreed to have brought the "worst winter weather for eighteen years".

I am not a climatologist, and would never claim to be, but the initial winter forecasts using some of the most powerful computers in the world were clearly unhelpful. Cynics would say, unjustifiably in my opinion, that if the experts can't forecast correctly for the season ahead how do they know what will happen in the next century.

Going back to autumn 2008 there were ominous signs that the weather was about to 'buck the trend'. October was on average about 1C colder than the long-term average although sunshine and rainfall were above expectations. After seeing the temperature hit the dizzy heights of 22.9C in Berkshire on the 12th there would be a sudden switch in the winds for the final week. A series of deep depressions and severe gales caused the abandonment of the Lake District marathon on the 25th with numerous participants having to be rescued in the worsening conditions. By the 28th a small but quite intense weather system of Arctic origins left Ireland and tracked across Wales towards southeast England bringing the heaviest October snowfall to the area for at least 80 years and possibly since 1880. Just one day later violent hailstorms and torrential rain led to serious flooding in the Otter valley, East Devon.

So a month of huge contrasts with that remarkably early autumn snowfall for southern Britain but there was to be more cold weather for the second half of November. Frequent heavy snow showers crashed into the north and east of Scotland along with northeast England on the 22nd and it was estimated that up to 30cm of snow was lying across parts of the North York Moors.

It was to be the coldest December for seven years as northerly winds continued to dominate for much of the time but it was probably just a typical winter month overall with no significant amounts of snowfall in the south. However, there was heavy snow at times in the Pennines, Cumbria and the Scottish mountains. After fresh snowfall in Lancashire overnight many schools north of Manchester were closed for a third consecutive day on the 4th. The final three days of the month stayed below freezing throughout at Kinloss in the Moray Firth.

January also began very cold with widespread overnight frosts and further snow for England and Scotland with sunnier skies across Wales. The next few weeks were stormier with gales and heavy driving rain for western and northern areas and it was cold enough for further snow in the hills. Overall it was the coldest January for about a decade and the final days of the month set the scene for the harshest winter weather yet as cold easterly winds swept in from the North Sea between high pressure over Scandinavia and developing low pressure over Biscay.

Without doubt the developments at the beginning of February will be talked about for many years by people living in the London area. By breakfast time on the 2nd there was 15cm of lying snow in parts of central London following frequent heavy snow showers overnight with parts of Surrey receiving considerably more. The first few days saw further snowfall progressing north across England into Northern Ireland and Scotland, although northwest Wales and the west of Scotland were better protected from the easterly winds and escaped the worst of the weather.

These were the heaviest snowfalls to hit such a wide area of southern Britain for a number of years and to help explain why it happened, and the primary cause, there is a need to understand what happens a few miles above street level. The ribbon of fierce winds, commonly known as the jet stream, is constantly changing its appearance as it flows around the earth and the weather we experience on the ground is largely governed by its progression. Sometimes the flow becomes 'blocked' and this can lead to a prolonged spell of quiet, anticyclonic weather across Britain. Alternatively, as was the case in early February, the jet stream moved south and split. The resulting 'cold pool' brought unusually cold air a few miles above us in the atmosphere, which was to react with the relative warmth of the sea off southern England.

The situation was further compounded by the suggestion of 'sudden stratospheric warming' much higher in the atmosphere during the second half of January which was thought to have influenced the jet stream lower down.

However, it was the movement of the 'cold pool' which led to the heaviest falls of snow due to the instability in the atmosphere. First it headed west across southern England, it then tracked south before moving east again close to northern France over the first six days of the month. Significant flooding in the West Country on the 9th was quickly replaced by heavy drifting snow around Dartmoor as temperatures plummeted again in the northerly winds. Further significant snowfalls affected eastern counties on the 12th, most noticeably in Northumberland and Angus.

Conditions eased thereafter as February returned to 'normal' but it was an unusually severe month for southern areas even though Scotland had experienced more typical winter weather. In fact the average temperature in Northern Scotland was about 1C above the norm!

For statistical purposes the year is divided into four equal parts so the winter of 2008-09 included the months from December 2008 to February 2009. Mean temperatures for the whole of the UK were 0.5C below average, provisionally making it the coldest winter since 1996/97 but for England and Wales it was the coldest since 1995/96. Despite the severe start to February milder conditions in the second half of the month resulted in near-normal temperatures overall. Rainfall amounts over the bulk of the UK were below the long-term average with Wales having its twelfth driest winter on records dating from 1914. Sunshine amounts were well above average during both December and January, although February was duller. Overall it was the eighth sunniest winter on record from 1929.

Is this a classic example of Mark Twain's phrase 'lies, damn lies and statistics' being justified? From a statistical point of view it could be that in fifty years time the winter of 2008-09 will scarcely be remembered, except for the events of 2nd February when London ground to a halt. Was this simply a 'one-off' occasion when the media compared the situation with winters of a bygone age knowing that they would become increasingly rare events if we believe what the experts tell us about global warming? Or maybe our perception of the weather is what really matters and how a succession of notable weather events in one 'slightly worse than average' winter are far more memorable and significant than statistics.

The extreme cold of January 2009 also brought some spectacular winter sunsets, as seen here in the Hope Valley in Derbyshire. (YPN)

# 1.

# A YORKSHIRE CHILDHOOD

I was born on the stroke of midday on 11th July 1952 at Halifax General Hospital on a typically cool West Yorkshire summer's day.

Food rationing was to last another two years following the end of the war and the latest census of Great Britain indicated that six per cent still had no piped water supply and thirty-seven per cent of households had no fixed bath. We were no exception!

What you got for your money in those days is scarcely believable now. A completely self-contained, unfurnished bachelor flat in London SW1 cost £80 per annum plus rates and a Georgian style detached residence five miles from Princes Risborough in Buckinghamshire with one acre of ground was on the market for £5,100. On the other hand the University of Hull was advertising for a lecturer in the Department of Philosophy and Psychology, annual salary £500.

After leaving hospital it was a twelve-mile journey from Halifax to Todmorden through Calderdale. It is believed that the name of my home-town derives from Tott-mer-den – 'the valley of Totta's marsh' – and there is little doubt that its location at the bottom of a steep, craggy valley has been both a lifeline and a hardship for generations living there throughout the centuries.

Home for me was Commercial Street, a modest row of houses sandwiched between the River Calder and the Rochdale Canal. It might sound semi-rural but down in the valley bottom this was 'Lowry country' with row upon row of terraced housing nestling amid the factories and mills of industrial West Yorkshire. My street consisted of three blocks, which all overlooked the river at the back, but looking across from the front door was a mill owned by the British Picker Company running virtually the whole length of the street. A small engineering factory and then stables stood further down before the street ended at the entrance of the Sandholme iron foundry.

George 'Sunny' Sunderland kept his horse 'Jock' in the stables when he wasn't out on his rounds delivering coal most days to some of the local factories. He was a wonderful old character who always had a cheery word for the kids on the street but was tragically killed in a road accident in 1962 shortly after he retired.

Allied to cotton production, there was a time when Todmorden had ninety per cent of the picker trade in the British Isles but those days were quickly disappearing. Still, the picker factory did at least provide us with a wall to

**When I was very young, aged 1 or 2 in 1954.**

smash a football against or chalk the wickets for a game of cricket. It was a relief when the cobbled stones were eventually replaced because it did make batting very tricky against the erratically moving ball!

Every house on the street in the fifties had their toilet in a stone 'lean-to' at the back and it would be another nine years or so before we had a bathroom built on the second floor of the house forcing me up into the extra bedroom cunningly constructed in the attic. Clearly this was a ploy by my mum and dad to get me closer to the weather!

What Commercial Street lacked in culture and open spaces was compensated for in warmth and a sense of being. It was a blissful existence shared by several other families like ours or elderly people who had many a tale to tell, seeing out their days in an environment they were clearly comfortable and happy with. There were enough kids of similar age and interests so we could have regular games of football or cricket, although my younger sister Susan was probably not all that keen on being forced to play goalkeeper when none of the lads fancied it!

In those early years our street was our playground, and our education, because it was a place full of character and history, both people and places. This was still a throwback to the industrial revolution and the pace of life was 'slow to steady'. Milk was still being delivered by horse and cart and the cobbled street was not replaced by tarmac until I was about eight years old.

What you've never had you don't miss and I was happy with my lot!

Holidays in those early days were always governed by 'wakes weeks' and Todmorden usually 'emptied' in the second and third weeks of July, about the time of my birthday. As part of the original Lancashire & Yorkshire

**Todmorden town centre in flood – about 1931.**

Railway it was easy to reach either the east or the west coast and my recollection is that for many years we either went from Platform One to holiday at Blackpool and Southport or from Platform Two to travel east to Scarborough. Obviously the weather could be poor on either coast and plenty of time was spent in the amusement arcades rather than on the beach. But I was told by my parents that we all played football on the beach at Scarborough one summer with Peter Taylor, then the Middlesbrough goalkeeper but more famously the long-time footballing partner of the legendary Brian Clough.

My first school was Castle Hill Infants down Halifax Road which I have bad memories about, simply because of the school dinners.Beetroot and mashed potatoes was not a culinary treat for me but it seemed to be a standing dish in those dim and distant times. I can also remember the foldaway beds we had to sleep on during the afternoons, whether we were tired or not. I'm sure kicking a ball around in the playground would have been more enjoyable so after just one year I was off to Roomfield Junior School, still on Halifax Road but closer to the town centre.

When I passed my 'Eleven Plus' in 1963 it was the opportunity to further

**Kids on the street in humble Commercial Street, Todmorden. I'm second left at the back.**

my education at Todmorden Grammar school, a proud co-educational establishment boasting two Nobel Prize winners. The first was Sir John Cockcroft (1897-1967), a renowned physicist who helped 'split the atom' with colleagues at the Cavendish Laboratory, Cambridge, and was awarded the Nobel Prize in 1951. Sir Geoffrey Wilkinson (1921-1996) achieved his award for Chemistry in 1973 whilst a serving Professor at Imperial College in London.

An everlasting memory of my early Grammar School days is, perhaps surprisingly, music lessons listening to 'Morning' from Peer Ghynt as the sun rose slowly over the Pennines.

It was a moving piece synonymous with breathtaking views of the dark peaks of Calderdale silhouetted against the golden globe with smoke swirling around the lines of terraced houses and factories in the valley below. This was a view which probably acted as a catalyst for a career in meteorology because it was the impact of the hills on the weather that stimulated my interest. The heavy snowfalls, violent thunderstorms and torrential downpours were all typical of the Pennine weather I grew up with.

In spite of the Lowry townscape, Todmorden was a contrasting world of rivers , streams, woods and open spaces of the hillside slopes, culminating in the 'tops' where reservoirs nestled in the natural hollows of the millstone

**The back of Commercial Street looking across the allotments.**

grit. This is Pennine Way country, just twenty-four miles from the southern extreme at Edale in Derbyshire on the way north to Kirk Yetholme in the Scottish Borders.

When not kicking a ball I would undertake numerous family walks with my mum, dad and Susan in this varied landscape. Rising from the valley bottom through craggy gorges dotted with wildflowers and ferns you reached the first plateau at about 300metres. Here could be found pasture where cows were kept to supply milk to the town below. It was a fond memory on a Sunday afternoon to call at one such farm, that of Leonard Barker, and recharge batteries with a cup of tea or orange juice and then set off home having bought a carton of fresh cream!

However, my mind also goes back to the time I picked up a wasp's nest believing it to be an old cricket ball. To understand what came next requires an understanding of my mother's fear of these crazy insects. Suffice to say my being stung three times on the same arm was nothing to the entertainment provided by her frantic gesticulations and shrieks as she tried to escape their attention.

Onward and upward to 450metres and Stoodley Pike, a marker for the Pennine Way and a stone monument built at the second attempt in 1854 to commemorate the surrender of Paris following the Napoleonic Wars almost two hundred years ago. The weather at this height can be wild; the sheep which punctuate the millstone peaks sought protection here from the bitter winds. Some sheep died in this refuge from the inhospitable winter weather and you ventured along this stretch of Langfield Common with a very close eye on the gathering grey skies which could shroud the hills in seconds.

It was with a sense of great pride that I stood beneath Stoodley Pike in 2000 recording a piece for Songs of Praise with presenter Kevin Woodford. The theme was Mother's Day and what my family life and home town had meant to me all those years earlier. I reminisced, reflected and emphasised the strength of character shown by Todmorden people through my lifetime with an overriding independent nature probably contributed by the position of the town on the border between Lancashire and Yorkshire. The boundary used to run beneath the Town Hall right in the centre so it was possible to stand with one foot in each county – there has for many years been a dichotomy as a result.

At the heart of the Pennines, industrial Todmorden was founded on cotton and its red rose leanings come because, although it is now firmly in the white rose county of Yorkshire, it has often been referred to as "the last Lancashire cotton town". Just five miles along Calderdale back towards Halifax you come to Hebden Bridge which owed its existence to wool. Like its neighbours, the conventional industries and original life support of the town have since died and it is now a fashionable enclave for artists.

John Fielden was part of the cotton dynasty for which Todmorden is justifiably proud and provided work for hundreds of men and women, especially in the nineteenth century. As MP for Oldham he later helped to implement the revolutionary Ten Hours Act restricting the number of hours

**Roomfield Junior, my now demolished primary school in Todmorden.**

worked by children in the mills. The Fielden family commissioned many fine buildings in the town, including the Town Hall, but John's eldest son, Samuel, was passionate about cricket and as the founder of organised cricket in the town he served as landlord, player, treasurer and president. After a couple of temporary grounds part of Samuel's recently acquired estate at Centre Vale was established as home of Todmorden Cricket Club around 1850.

It would be over a century before I would be taken to the cricket ground, first in a pram and later as a junior member before eventually playing at various levels having transferred my passion for the game on the street 'to the middle'. I spent hours in the early 1960s with friends scoring in the 'cheap seats' on the boundary edge. Famous cricketers, some already playing Test matches and some about to break through, improved their game by playing as paid professionals in the Lancashire League (Todmorden CC was the only Yorkshire team in the League). The earliest I remember from the West Indies would be Seymour Nurse, Conrad Hunte and Wes Hall with Neil Hawke and Ian Chappell from Australia. In 1961 my home club employed the great English fast bowler Frank Tyson and his speed was awesome.

On an amateur level two future Test players began their careers at Todmorden, both of whom I believe also preceded me at the Grammar School. Derek Shackleton headed south to Hampshire and played seven times for England from 1950 to 1963 and Peter Lever joined the 'old enemy' at Old Trafford playing seventeen Tests from 1970 to 1975.

I have to say my achievements for Todmorden CC were extremely modest in comparison, although I always 'talked a good game'! The junior Third team played twenty overs a side games on weekday evenings against other Lancashire League teams such as Ramsbottom and Rawtenstall. I grew in confidence after making my debut at fifteen, culminating with a timely 50 to win the game on my eighteenth birthday! Some of that ability also made me a regular in the Second team for a couple of years but the only worthwhile performances I could manage were a top score of 22 against Burnley and best bowling of 4-24 against Bacup. Modest indeed!

Stanley Hollows was captain for most of my games but he had played much of his career in the first team before dropping down a level to encourage the next generation. Among other stories I vividly remember him telling his aspiring youngsters was the time he had to face Charlie Griffith on a hat-trick. Now Charlie was a West Indies Test cricketer who had been accused of 'chucking' during his career but was still in his prime in 1964 when acting as professional for Burnley. Unsurprisingly he wreaked havoc most weekends and through the season captured a total of 144 wickets in just 26 league games, some of which would have been hit by the weather.

As Stanley left the pavilion to face the hat-trick ball he realised he was without that most vital of accessories – his box! Trepidation had already turned to fear as he walked across the boundary rope but now he had to be both the hero and above all preserve his manhood.

Needless to say Charlie got his hat-trick that day and Stanley turned away intact.

For once he didn't care about being the hero and was more than happy to have become another victim in an unforgettable season for the West Indian!

Only once was I selected for the First team, in truth during the summer holiday period when a few regulars were missing, but ironically the game never started as the rain teamed down all day. There was probably some divine intervention involved for which I was eternally grateful as I hadn't played for about a month.

Horse racing was a further sideline and distraction throughout my schooldays – indeed for most of my adult life as well – and I blame my ancestral line solely for that particular pastime!

Until I began studying my family tree in the mid-1980s I had always believed that the Kettley family was from Yorkshire going back through the generations. There was no doubt that my mum (Marian) and dad (Harold) were both Todmorden born and I also knew that my grandfather, Fred, was born near Goole on 2nd March 1882. He was one of twlve children born between 1873 and 1903, although three never reached adulthood. Their mother was Mary Ann Kettley (nee Sheppard) who was also born at Barmby on the Marsh, East Yorkshire, on 3rd December 1852. She died in Todmorden on 5th January 1913 and is buried with three of her children in the graveyard of the parish church, adjacent to the cricket field.

To my astonishment my great grandfather, Walter James Kettley, was not from Yorkshire at all but was born at Blackmore near Ongar, Essex, in 1853. As I understand it he was born or brought up in a pub called the Black Bull in the village – not a bad way to start your life and presumably he has passed his liking of real ale down at least three generations already!

So great grandfather was an 'Essex boy' – I can hardly believe it – but even more interesting he was also an apprentice jockey in the early days before eventually spending most of his working life as a groom.So it could be that he met my great grandmother when he was racing up in Yorkshire as a jockey because they had their first child when they were both about twenty.

Clearly, then, horse racing is in the blood and it is a passion that has remained with me and even become part of my consultancy business in more recent years – but more of that later.

I know it was customary for my dad to have a small bet at work on a Friday afternoon with a few of his workmates at the shuttle factory where he worked much of his life. My first tiny footsteps into the world of gambling were as a teenager on a Saturday afternoon when I would pick out four horses which I hoped would win. The 6d 'yankee' cost me 5s 6d, just a small investment but like millions of others I would live the dream for a few hours that my doubles, trebles and accumulator would bring me a small fortune, but of course it never did.

The Grand National at Aintree is arguably the biggest racing event on the calendar which captures the imagination of the British public, and for maybe just one day a year tempts them to part with a couple of pounds. The first I remember was the 1960 National, won by Merryman II, which I watched with the rest of the family on our tiny black and white television. Not for me a betting event at that time but a magnificent sporting spectacle which continues to capture my attention to this day, helped of course by the excellent BBC production skills and covered for many years by my friend Julian Wilson.

It would be about eight years later that I first went racing for real, or at least that was the plan. Along with John Walter, a pal of mine at school, we got a train to Castleford where we would take a fish and chip lunch in the centre of town before walking the two miles to Pontefract racecourse. But the 'fickle finger of fate' conspired against us because a notice board in Castleford bus station gave us the news neither of us wanted to read: 'Racing OFF – waterlogged'.

Maybe the weather was conspiring against us for a reason – but we were not happy – and we went home again with less money than we started with but not having had a single bet.

A few weeks later we repeated the mission with much more success and in a moment of madness we organised ourselves a 'racing holiday' in the summer of 1970. Having finally left school we studied the racing calendar and implemented a ten day journey which would include two racedays at Catterick Bridge, the middle Saturday at Ripon for the Great St Wilfrid Handicap and then the three days at York for the magnificent Ebor Meeting. The bookies were safe as we travelled from one meeting to another because our stakes were understandably small, so much so that a one shilling win bet on 4/9 chance was met with a cry of "What are you trying to do, break me?" from one twitchy bookmaker.

Enjoying an evening meal and a couple of beers in a city pub after the first day of racing in York we entered into conversation with two friendly bookies who offered us work on the middle day of the three-day meeting. The task was simple and the experience fascinating as we spent the afternoon walking along the line of the other bookmakers, noting their prices and reporting back to our new 'friends' if they were adrift with their odds. It was a day with a difference and we were both rewarded with our small wage as they packed their bags at the end of racing. At least we had an extra pound in our pockets for the final day of our 'holiday' on the Thursday.

**Horse racing gave me an opportunity to combine photography with my passion for the Turf.**

First and foremost it's the spectacle of horse racing I adore and it allowed me the opportunity to combine another hobby, namely photography, with my passion for the Turf. On most racecourses you can wander about on the inside of the course and get a little closer to the action away from the bustle of the betting ring and paddock. Standing by the fence as the pace of a race increases you get a sense of the adrenalin rush all jockeys must feel. To capture the moment the combined force of horse and rider surge over a fence, and hopefully make it to the other side intact, is a moment to enjoy on camera. What you cannot feel on the still photograph is the rumble of the ground and increasing crescendo of noise as packing horses approach the obstacle but that can be made up for with the expressions on the face of horse and jockey. Fear, excitement, will to survive and anticipation of what lies ahead are all factors.

Surprisingly I have never sat on a horse in my life, despite my genes, but I feel I could still enjoy the experience with some gentle persuasion and a kind conveyance. The power of a horse as it gallops down the straight mile at Royal Ascot or Newmarket is breathtaking and I admire the riding skills of the jockeys enormously, but 'flat' racing for me remains secondary to the sheer power and spectacle of jump racing. Those jockeys deserve a medal every time they sit on a horse and what they achieve in their chosen career is well earned and a testament to their courage.

My chosen profession did not start out in quite the same way as you would imagine and it was during my time at Todmorden Grammar School that the first seeds of a career in meteorology were sown. I had always favoured the sciences rather than the arts, although I would be the first to admit at a modest level, despite the high quality of teaching at the school. By the sixth form I had listened to several guest speakers who chose to talk on their own careers but I was never inspired sufficiently to deviate from my 'hobby' of the weather. Now I have to say that Todmorden Grammar School expected sixth formers to leave with good 'A' Levels and continue on to university.

Presumably their philosophy was that you can't have too many Nobel Prize winners!

I was of the opinion that it would be nice to choose a career, get a job and have a break from education, at least for a while. It was then radical thinking but the 'gap years' have become much more fashionable since the time I was leaving school. At a careers evening in the school library we were all invited to speak with the representatives from the local education authority on our intended plans for the future, be it educational or otherwise. For my part I was delighted to get the information I needed to join the Met Office via the Civil Service Commission. I also spoke with another official about the possibility of becoming a sports journalist, mainly because I was interested in many sports albeit fairly useless at most. My questions were met with blank faces and unanimously negative comments like:

"Oh no, you couldn't pursue that line because you're not even doing English 'A' Level – out of the question."

"Have you thought about teaching?"

I knew it was time to leave.

Two referees were needed in order to apply for a Civil Service post and I am eternally grateful to Mrs Alice Crossley, the wife of a businessman in the town, and also PC Alan Bolton who was lodging on Commercial Street at the time. I was then invited for an interview at the Manchester Weather Centre, Royal Exchange, on Monday 20th April 1970 at 9.30am. A brief tour was followed by a chat with the boss and then a written test which all proved very successful.

A subsequent offer to become an unestablished scientific assistant was sent from Met Office HQ, Bracknell on 29th April – signed JA Gregson, Your Obedient Servant. Finally an offer of employment on an initial trial period of six months was to commence at Manchester Airport, 24th August 1970. Arrive 09.30 or as soon after as possible.

I achieved my Applied Maths and Physics 'A' Levels at the required grades and my school days were over. Now I had a career, although nobody could have imagined at the time how it would evolve.

Media – you're not even doing 'A' Level English!

## 2.

# WINTER 1962-63

I didn't know it at the time but the severe winter of 1962-63 was probably the main reason why my future career developed as it did. It turned out to be the coldest and snowiest winter in the south since 1740. Cars could be driven across the Thames. Pack ice formed a quarter of a mile outside Whitstable Harbour. One family was marooned on a Dartmoor farm for sixty-six days. At least forty-nine people died ... and for a shy ten year old growing up on Commercial Street in Todmorden, West Yorkshire these were changing times.

For those of us with a passion for football there was no way of expending our energy as the playground remained ice-bound for a full ten weeks. Looking back 'every cloud has a silver lining' and as I was about to take my Eleven Plus exam the almost complete absence of any outside recreation did at least concentrate my mind on academia. If I passed the exam I would be offered a place at the Grammar School just a mile away from home but the alternative comprehensive school was at Mytholmroyd, a seven mile bus ride away. The third alternative was Scaitcliffe Secondary Modern, also in Todmorden, but never considered as an option for those with loftier ambitions and was more often a vehicle for apprenticeships in local business. At ten I had no idea what those loftier ambitions would be but I knew that an office job was my likeliest route forward.

Meteorologically there had been indications of what was in store during the late autumn of 1962, many weeks before Christmas. On 8th November temperatures began to fall across the British Isles and just one week later the severity of the cold was enough for heavy snow to fall which began to drift in the gale force winds throughout the West Country.

Other perennial problems characteristic of 'old-fashioned winters' were evident by 3rd December when freezing fog became widespread in many areas, not least at London airport where many flights were to be diverted and outgoing flights grounded.The next few days brought worsening conditions reminiscent of the days prior to the passing of the Clean Air Act in 1956. Smog – a yellow, toxic cocktail of sulphur dioxide and smoke from coal fires – returned to the streets of towns and cities and conditions were described as the worst since the 'Great Smog' of 1952. All means of transport, including shipping at the Port of London and the Clyde, was at a standstill due to the worsening conditions and the death toll from respiratory illnesses escalated. Several leading London stores complained that the fog was keeping Christmas shoppers away, with turnover well down.

London airport re-opened on the 7th after four days of closure offering 'a light at the end of the tunnel' for most areas of Britain. Normal service was resumed for over two weeks until the arrival of the Christmas festivities and then the country embarked on what for a generation of people, including me, was to be the worst winter ever seen. Britain virtually ground to a halt on several occasions during that severe winter and as the weeks went by it felt there was no end in sight. Was this the end of civilisation as we knew it?

The severity of the frost and the increasing disruption to the country's infrastructure, not to mention the sporting calendar, was most acutely felt between Boxing Day and early March.

The first heavy snowfall had arrived in northern Scotland on Christmas Eve with the passage of a cold front which turned the rain into snow as temperatures tumbled. It was this blanket of snow which then gave Glasgow its first white Christmas since 1938, although sleet had fallen in 1956. Strictly speaking the bookmakers' definition these days requires sleet or snow to fall on Christmas Day itself to qualify for a 'payout'. Perversely if the snow has already fallen previously and is lying 'deep and crisp and even' on the big day itself it would not qualify as a white Christmas. In the case of Glasgow, and many other parts of the country, 1938 was the snowiest Christmas Day prior to 1962.

Snow fell throughout the country on Boxing Day leaving roads "exceptionally dangerous" according to reports at the time, apart from the far southwest of England. Overnight temperatures had plummeted to minus 9C in Cardiff and minus 11C in Birmingham.

The first signs that the sporting calendar was about to go into 'meltdown' was the postponement of nineteen football league matches with three also abandoned once they were underway. Horse racing was wiped out and it was cold enough for Poole Harbour to freeze for only the second time in twenty-five years. On the 27th further heavy snowfall led to major travel disruption with up to ten inches of snow on the runway at Gatwick. Rail delays made thousands of Londoners late as they returned to work after the Christmas break.

By this time I distinctly remember the ever-increasing depth of virgin snow lying the length of Commercial Street; only rarely did anyone dare to venture out in their Austin A40 or Morris 1000 traveller. More disconcerting was the weight of snow piling-up on the terraced roofs although from time to time you would hear the rumbling sound of snow crashing down onto the pavement below. It was safer to walk along the kerb edge rather than risk the equivalent of a wheelbarrow full of snow falling on your head. But it was all very exciting for us kids on the street, knowing that school would not be open again for at least another week and this was as close as it got to 'heaven's playground'.

Back on the sporting theme another twenty-six football league matches were postponed on 29th December and the worsening travel conditions led to the death of ten people in snowstorms on the last two days of the year with people in Devon and Cornwall already comparing the severity of the snow to the winter of 1947. The strong winds were producing deep snowdrifts across southern England and by now thousands of miles of roads were affected by the snow.

There was heavy snowfall in the north too. Villages were cut off, roads and railways blocked, telephone wires and power cables brought down. Thousands of sheep, ponies and cattle were buried under the snow but several people died as well. A 65-year old railway worker, John Warren of South Benfleet, was killed by a train as he tried to clear the snow and ice from frozen points. William Starkey, a 60-year old milk roundsman collapsed and died while making deliveries in the snow and a husband and wife, Arthur and Daisy Barber, were found dead in their car after being trapped by drifts on Osmington Hill, Weymouth. Their daughter, Mrs Sheila Reed, her seven-year old son Ian, and Thomas Cumbes of Dorchester, who were trapped with them, all survived.

Just nine miles away from my home town, gales brought down a 200ft chimney in Rochdale but there were also signs of severe industrial unrest as power workers decided to ban overtime and work to rule. This would now bring an already serious situation to a head with unparalleled repercussions across Britain.

On a lighter note the Met Office was now confirming that seven towns had broken their December sunshine record of 100 hours which had existed since 1917!

By New Year's Day virtually the whole of the UK was covered in snow and, despite the best efforts of the army, deep snowdrifts were now common across the West Country with Weymouth and Bridport completely cut off. There were countless reports of heroism and rescues of both people and animals all across the country with abandoned vehicles littering the road network. Continuing blizzards over the south of England were now said to be the worst since 18th January 1881, an occasion when even Oxford Circus in London had been covered in a fifteen foot snowdrift.

I wonder if Boris Johnson, the current Mayor of London, would attempt to cross London on his bike in those conditions as he did at the start of February 2009 when snow hit the capital. At the time we heard of the dwindling stocks of grit and salt which are used to help clear the roads but back in 1963 up to five hundred lorries from all parts of the country were queuing for rock salt at a mine in Cheshire as conditions worsened.

The economic consequences by early January 1963 were becoming devastating. Thousands of gallons of milk were poured away by West Country farmers who had been isolated by the snowdrifts. Sussex farmers were also forced to pour away their milk and 10,000 gallons were at one time awaiting collection on the roadside between Mark Cross and Wadhurst. Hardship payments were made by the Milk Marketing Board but many householders were without their daily delivery of milk – these were the days long before supermarkets became the main source of your 'pinta'.

Farmers in the Fens and East Anglia who had already lost thousands of pounds worth of celery because of the frost, were fighting to save 175,000 tons of sugar beet – valued at £1million – which was frozen into the ground. The Cantley sugar beet factory faced closure if supplies of beet did not arrive soon.

The worsening conditions prompted the RAC to send a report to the Ministry of Transport outlining the use of road-heating appliances to reduce ice hazards. An internal committee at the RAC had been in existence for over a year looking at this possibility but apparently it was merely coincidence that they sent their report as road conditions became intolerable.They recognised that under-surface heating of roads would be too expensive for local authorities to implement. However they highlighted sixty-five danger points, mostly at intersections and on hills, at which road-heating might improve the safety of road travel. We are still awaiting this development almost 50 years later!

The railways were perilous too. The three-man crew of a 600-ton coal train jumped to safety into a snowdrift only seconds before the train crashed at Bradford after sliding through a set of points on the icy rails. The runaway engine "ploughed into a concrete stop-block, leapt into the air, shot thirty yards along a platform, wrecked a steel water tank and stopped perched over a thirty-foot drop".

There was a startling variety of weather across the country with just one thing in common – all of it was bad!

There was prolonged snow on Friday 4th January over the north of England, Wales and most of Scotland but a slight thaw further south helped to ease travel on some roads. Meteorologically a classical set-up had become established over the past seven days which had ensured heavy snowfall in many areas of Britain. Low-pressure rain-bearing systems over the Bay of Biscay were in conflict with the anticyclone or high pressure belt stretched between Greenland and western Russia. The air was already bitingly cold across much of Scandinavia so the strong, cold easterly winds sandwiched between the two zones of weather left us in the battle zone. The weather from the south was essentially 'blocked' by the high pressure so we experienced the increasingly frequent bands of heavy and persistent snow. Drifting of the heavy snow occurred as a result of the strong winds from the east and the fact that temperatures were so cold.

Todmorden lies in the Calder valley where the A646 between Burnley and Halifax meets the A58 to Rochdale – a natural route centre where the

Temperatures in parts of Britain fell to minus 22C in the second week of January 1963, with snow already on the ground continuing to freeze hard at night. This snowplough is struggling to clear the road across the Pennines from Reeth to Brough. (YPN)

railway, road, river and canal all converge. The M62 did not exist in 1963 and, although there were several road links across the Pennines, the road through Todmorden was the only valley route. By this time the higher routes were at the mercy of the elements and had become impassable leaving Todmorden gridlocked and headline news! For me this was very exciting because it felt as though our town had just been discovered with television crews lining the main road. Under normal circumstances we never appeared on the television, although I do remember the BBC Home Service recording a radio episode of 'Down Your Way' about this time, presented by Franklin Engelmann.

With the snow piled up by the side of the main road the image was one of cars, trucks and lorries crawling through the narrow valley. No-one was going anywhere in a hurry and it is true to say you could walk faster, although even walking was a precarious exercise with fresh snow falling on icy pavements. There was a feeling of huge anticipation watching the early evening news to see the miles of queuing traffic crawling through Todmorden – and was I captured in the background making a cameo appearance on national television? To my knowledge I never was but the affects of the weather on the infrastructure of my small part of northern England must have been slowly impregnating my mind.

Fairly recently a heart-warming story came to life encapsulating the severity of the winter weather in early 1963. Christine Fairbank told her horrifying story when, as a nine year old, she and her mother nearly died in a blizzard on the moors above Rishworth, just a few miles from Todmorden. Christine's father, Albert, was the keeper at Green Withens reservoir near Windy Hill, close to where exit 22 of the M62 is now. She recalls being out with her mum one morning in Rishworth and then catching the Oldham bus as far back as Windy Hill.  It was cold and sunny at the time but as they began their one-mile walk from the bus stop to their home near the reservoir the weather closed in. At nearly 400metres above sea level the weather can change almost without warning and before long the snow was falling heavily. They battled against the blizzard on exposed moorland dangerously close to a deep-water conduit:

"The gale-force winds blew straight into our faces, whipping up ice and peppering us hard. It was very painful and each gasp of breath became more painful.  I cried with the pain and tried to shelter behind my mum and under her coat, but we had no mercy from the dreadful conditions. It took us two hours to reach home but then we could not move our hands or fingers to grip the key to open the locked door.  There was nothing more we could do to help ourselves."

Fortunately the pair had been spotted getting off the bus by a surveyor who was working on the moors collecting data in connection with the motorway that was about to be built across the Pennines. Christine picked up the story:

"When he completed his work he returned to see if we were all right.  He found us huddled together on the doorstep like frozen statues.  I definitely owe my life to the M62 motorway!"

The surveyor was John Shakespeare, a 24-year old man living with his wife and son in nearby Mill Bank.  Forty-three years later, and still living in the area, he read Christine's story in the *Halifax Courier* and decided to get in touch. Back in 1963 his workplace was a small cabin at Scammonden. Not any old cabin but a place where, coincidentally for me, he collected readings from a number of weather stations in order to help the motorway engineers in the design of the motorway. He had been told not to go out in a blizzard and even recalled the death of another reservoir keeper during that infamous winter.  John recalls seeing Christine and her mother get off the bus and that the weather was deteriorating.  When he finished his work he took the route to their home in his four-wheel-drive Austin Gipsy, found the pair and got them inside.

Across the rest of Britain the second week in January was dominated by high pressure and so the snow already on the ground continued to freeze hard at night although there was a slow thaw in the south by day.  The temperature fell to minus 22C at Grantown-on-Spey, Morayshire, on 7th January and there were reports that sheep were being eaten alive by foxes on Dartmoor.  Two days later the chief operations engineer for the Central Electricity Generating Board said that the London area survived its peak demand for power "by the skin of its teeth." On 11th January shop stewards at the London power stations voted for a "more rigid application" of the work-to-rule on a day when there was reduced voltage throughout the country.  Bristol Harbour was now frozen and by 14th January ice covered the sea for a mile along the front at Herne Bay, Kent, and extended at least

**The River Thames frozen over at Windsor Bridge for the first time since 1897. A cyclist braves the ice on 23 January 1963 – the coldest night of the winter. (PA)**

100 metres from the promenade.

The second half of the month saw further blizzards, initially across Scotland and the Northern Pennines but then further south across the rest of England and Wales. By 20th January snowploughs were unable to cope with the fresh falls of heavy, drifting snow and it was said that half the main roads of southern England were again impassable. Thousands of cars were abandoned, including many on the M1, and other main routes out of London such as the A1, A5 and A6 were forced to close. Southeast England had now been in the grips of the big freeze for about four weeks and even the Thames froze over at Windsor Bridge, the first time since 1897.

Cross-Channel ferries were suspended due to ice floes off the French coast and all air traffic was diverted from London Airport because of snowdrifts on the runway. The railways didn't escape either with numerous reports of trains becoming stuck in snowdrifts and locomotives in the Western Region even freezing-up while they were running. Remember these were the pre-Beeching days when branch lines littered the entire network.

The use of helicopters was increasingly vital in these times of severe wintry weather and 283 workers were evacuated from the snow-bound early warning station at Fylingdales in North Yorkshire, high on the North York moors. At the time it was said to be the biggest helicopter airlift ever carried out and was completed successfully with savage winds gusting up to 80mph.

Four helicopters were used to remove the staff who had been trapped for five days with fuel supplies running low.

Two climbers were killed near Oldham on the 450metre high Wilderness Crag. In all four members of the Gritstone Climbing Club had been caught by the avalanche of snow and rocks resulting from the exceptionally severe winter weather. Two were overwhelmed by the avalanche, leaving the other two to seek help from a mountain rescue unit, police, ambulance men and troops from Preston and Ashton-under-Lyne. In all more than one hundred men took part in the rescue operation but the stricken pair were finally dug out from beneath the snow on the morning of Monday 21st January. Another walker died at nearby Ramsbottom and yet another man was found dead in a stranded car near Blackburn.

All three incidents were within twenty miles of Todmorden where my own schoolyard was now thick with layers of ice and snow. No chance now of a game of football, indeed it was a challenge to even stand up in these treacherous conditions. The daily walk to school, probably about half a mile each way, was precarious to say the least with the main A646 again gridlocked with crawling traffic. Despite the police putting "every available man on point duty" there was a fifteen-mile traffic-jam in Halifax as thousands of vehicles again made for the only road open into Lancashire. It was said that the twelve mile journey between Halifax and Todmorden was taking five hours to accomplish!

It was around this time that the cold had become so intense that the Rochdale Canal running through the town had also frozen over. Used extensively for transporting coal and raw materials to and from the cotton mills between 1804 and 1937 its use had become simply recreational, mostly by fishermen. But now the wintry landscape meant the canal could be used for other sports, including football, although I never would have been so brave (or stupid) myself.

Eight miles down the Calder valley at Sowerby Bridge the canal ran behind the Commercial Inn on Wharf Street, run by Hubert and Doris Davaney. As the pub closed on yet another bitterly cold Sunday afternoon in January the assembled clientele decided to have an impromptu game of football on the frozen canal. Apparently they wouldn't allow their son Stephen, 13, to go on the ice for obvious safety reasons and I really don't blame them.

Professional football had of course been virtually wiped out by the severe wintry weather and the Pools Promoters' Association decided that drastic measures were required. Even when football matches were postponed there would now be a panel of experts to produce a hypothetical result. We were now witnessing the birth of the Pools Panel. They first met to predict the results of the postponed league programme of 26th January providing a wealth of football knowledge at professional level. Under the chairmanship of Lord Brabazon of Tara there was former referee Arthur Ellis, and retired players Tom Finney, Tommy Lawton, Ted Drake and George Young. Each was paid the princely sum of £100 but were kept in splendid isolation at a top London hotel to allow their deliberations to be carried out in strictest privacy.

It should be remembered that in those distant days the football pools industry was reckoned to have annual receipts of some £80 million so the cost of running the Pools Panel was trivial. As the winter freeze progressed so the panel was used on a regular basis and millions continued to pit their wits against this esteemed group of former professionals in the hope of becoming a Pools winner. The public was happy and so was the industry but inevitably eyebrows were raised on Saturday afternoons when their decisions were announced. Each match might have been 'only a game' but for many it was also the flimsiest of lifelines to security for life.

Prior to the instigation of the pools panel at the end of January the FA Cup Third Round had been decimated back on the 5th of the month with only three out of thirty-two matches played. It would not be until 11th March that all the matches were completed, a period of sixty-six days involving a massive 261 postponements. One of the matches postponed involved my beloved Burnley FC who had been drawn away to Tottenham Hotspur at White Hart Lane and was a repeat of the Cup Final won by Spurs just eight months earlier. The power situation in the country was so bad that clubs were instructed to re-arrange their matches on a weekday afternoon to avoid the use of floodlights. So the match went ahead on Wednesday 16th January on a treacherous pitch with fresh snow lying on top of frozen ground.

I was ten at the time and still at Roomfield Junior school but I distinctly remember the afternoon of the game. My class was timetabled for 'music and movement' – a quaint double lesson to end the day involving prancing about with hoops and doing bizarre things with bean bags! The headmaster, Harry Wilson, was a devoted Burnley supporter and he was so excited to hear that we had won the match in London 3-0 that he was going round the whole school announcing the result. I could understand his pleasure and the result was certainly not lost on me because it was a big shock, perhaps only surpassed by my hitherto strict headmaster, who often brandished a cane (as they did in those days), almost skipping through the hall!

January 23rd was generally the coldest night of the winter and the British Insurance Association estimated that already the weather had cost more than £5 million in claims. The following day there was further trouble on the railways as diesel fuel, points, coal and water troughs froze and one train from London to Manchester took virtually twelve hours to complete the 189 miles. But they were the lucky ones as many trains didn't run at

**The beginning of February 1963 saw the West Country bearing the brunt of all that nature could throw at it. Floodwater on this field at Thorverton, near Exeter, changed overnight into ice floes up to three feet thick. (PA Archive)**

all. Bonfires were lit in the streets of Paddington to prevent water freezing in the stand-pipes. These were now desperate times and the Cabinet met at the House of Commons to discuss emergency measures.

A slight thaw on 24th & 25th January in Scotland and northern England sparked avalanches on the A57 Snake Pass between Manchester and Sheffield. Blasting operations failed to clear the snow and rock debris so it was decided to keep the pass closed for another fourteen days. The weekend of the 26th & 27th coincided with the worst power failure in the national grid in thirty-five years of operation. Ironically the weather was slightly less cold by then and the slight thaw continued allowing the National Coal Board to get supplies moving again. Eighty shiploads of coal left the northeast for London and many other parts of southern England while 4,000 lorries worked through the weekend moving coal.

The temperatures at Kew never got above freezing on sixteen days throughout January and the highest temperature recorded there that month was a miserly 2.7C, the lowest overnight being minus 10C. Many parts of southeast England constantly remained blow freezing from the 16th to 25th! It was to be the coldest month of the century!

By 4th February another blizzard brought further chaos to road and rail travel with Wales and southwest England again bearing the brunt of all that nature could throw at it. There was a huge reservoir of extremely cold air now anchored from Scandinavia to the west of Ireland and where low pressure systems attempted to force their way into Britain the inevitable result was another bout of heavy snow accompanied by strong winds. A train became marooned on the edge of Dartmoor with fifty passengers on board and seventy lorry drivers took refuge in a school between Exeter and Okehampton after being surrounded by deep drifts.

Unusually, parts of Scotland had not seen the worst of the winter weather but on this occasion they were not immune from the blizzards. Two school buses were stuck in Midlothian and it was left to local farmers to rescue the children but 150 lorry drivers were also caught between Lanark and Abington and they spent the night in a public hall. Conditions continued to worsen over much of Scotland and northern England with reports of nearly 1,000 vehicles stuck in snow on the A1 near Alnwick and severe drifting meant that even Edinburgh became cut off.

Over in Northern Ireland helicopters flew food supplies to villages isolated in County Londonderry and at Belfast Airport staff were marooned for the night.

A ridge of high pressure in Germany and a deep low pressure system off Ireland spawned further strong winds and bands of snow on the strong south-easterly winds but those same winds were absorbing slightly warmer air in Devon and Cornwall. As a thaw set in there was now an increased threat of flooding in the West Country and police had to dynamite ice on the River Exe to prevent such an event. The thaw spread north across central areas of the UK very briefly before more snow returned to most of England and Wales.

*The Todmorden News and Advertiser* led with a story in their weekly newspaper on Friday 8th February about how schools were coping with the freeze and also how the 'false dawn' of an impending thaw was quickly replaced by more heavy snowfall. On the previous Tuesday night biting cold gale force winds whipped the existing snow from the higher ground above the town and swept it back onto all the minor roads which had just been cleared of snow. So Wednesday morning broke with worsening conditions yet again with multiple road closures between Lancashire and Yorkshire. Traffic built up through the town along the A646 and A58 so by mid-morning long queues of lorries and cars were again stretched along Halifax Road and Rochdale Road from the bottleneck at the Town Hall in the centre.

As the harsh winter weather continued to bring huge volumes of traffic through the valley so the condition of the road surface was now deteriorating and becoming a major concern.

Residents in the town were also suffering from all kinds of hardship not seen since the snowiest winter back in 1947. Many homes were without water due to burst pipes or frozen mains and as each day passed local plumbers were having to deal with a lengthening list of emergencies. Staff from the Gas Board were also working round the clock trying to cope with the escalating problems associated with severe wintry conditions.

It was estimated that the cost of repairing the frost damage in the town was already running into many thousands of pounds. But out of adversity came plenty of goodwill, notably in the form of 'good Samaritans' from Dobroyd Castle School, an approved school run by the Home Office. According to the local newspaper the lads called on local residents and offered to carry buckets of water from hillside wells. One unfortunate householder, who couldn't provide a suitable bucket of her own, "was astounded when an hour or so later they came back with their own buckets from the school full of spring water. I've been able to wash clothes for the first time in three weeks," she said.

I did wonder though why they didn't simply just melt the fresh snow!

At the monthly meeting of the Calderdale Divisional Education Executive Committee a detailed report was given of the activities of the school during the severe wintry weather. Tributes were paid to the caretakers. The Divisional Education Officer pointed out that "there had been a great deal of trouble in the schools with bursts pipes, radiators and the heating arrangements. The caretakers in the division had done a remarkable job during the difficult period. They had borne a heavy burden in overcoming the water shortage".

At Heptonstall County Primary School, about six miles away but on higher ground above Hebden Bridge, "there had been no water since Christmas and the caretaker had been bringing water from his home and nearby houses to flush the toilets. The women who dealt with the school meals had also been bringing hot water from their homes for washing purposes."

In Todmorden itself "there had been no closures but the flushing systems of outside toilets had been frozen up." The report praised everyone and believed that "compared with the rest of the country they had been able to keep their schools open remarkably well, and it was due in no small measure to the staff."

I am sure that my school must have suffered like the rest from the problems of ice and snow although nothing stands out, except the hazards of the hard-packed snow on the frozen playground. Many a tumble was taken as the weeks went by but the question on everyone's lips was "When would it all go away?"

**A silent and deserted Snake Pass on 8th February 1963. The trunk road, linking Manchester and Sheffield, had been closed for over a fortnight owing to the threat of an avalanche. It was finally reopened the following day when gelignite was used to blast away the overhanging snow. (YPN)**

It came as no surprise to learn from the General Post Office (GPO) that the demand for weather information this winter was three times greater than usual.

On 9th February a team of quarrymen from Derbyshire County Council climbed 700ft up the mountain overlooking Snake Pass and used gelignite to blast away a 25ft overhang of snow threatening the road below. Giant bulldozers and a snow-blower then moved in to clear the road and finally reopen the pass.

The following day bulldozers struggled through snowdrifts to relieve five farms in Carmarthenshire following another heavy fall of snow in Wales and southern England.

The Meteorological Office, my employers for thirty years from 1970 to 2000, prophesised on 13th February that temperatures would shortly rise in all parts of Britain and that meant Devon's flood-emergency plan would be re-introduced. Devon got its floods the following day, along with other parts of the West Country, leaving many vehicles stranded but this time in deep water rather than snowdrifts. Further afield it was still snowing in Derbyshire and also Fylingdales where workers were marooned yet again.

Clearly this was another thaw which flattered to deceive as the high pressure from Iceland to Scandinavia began to draw very cold air towards eastern Britain. Fresh snowfalls reached Kent to Scotland by mid-month and led to the worst conditions of the entire winter between Perth and Inverness where cars were buried beneath fifteen-foot drifts. It would be several days before conditions improved in eastern Scotland and the story of an AA patrolman leaning on his shovel in the snow on the roof of a car buried beneath his feet is legendary.

At the House of Commons some MPs now decided that the Air Ministry had a lot to answer for because of their use of Celsius rather than Fahrenheit in weather forecasts. There were complaints about public "disgust, unhappiness and inability to follow temperatures given in Centigrade." Surely we hadn't reached the silly season already!

Over southern Britain there was a general improvement and for the first time since Christmas all main roads across Dartmoor were now open to traffic and more than half the scheduled football league matches were played on Saturday 23rd February. High pressure was dominating the weather charts at the end of February but now the wind was drawing milder air from further south in Europe and the worst was over, at least from ice and snow!

February had not been as constantly cold as the previous month but the period from 1st December to 28th February had seen the most severe winter for many areas since 1814 and for southern England the most persistently snowy since the eighteenth century. Snow fell somewhere in the country on twenty of the first twenty-three days of the month, and County Durham had a frost on every single night.

Meanwhile the extreme west of Scotland had been almost completely free of snow with remarkably contrasting conditions being experienced on the Isle of Skye. Here nearly 100 hours of sunshine was enjoyed in January followed by the driest February for thirty years. On the downside the winds were often strong and the frosts severe resulting in the blackbird and thrush population suffering as never before. The desiccating winds also led to three days of disastrous heath fires at the end of February killing livestock and burning thirty square miles of grazing land in the south of the island alone.

On 2nd March troops eventually relieved a farm on Dartmoor which had been cut-off by twenty-foot snowdrifts for sixty-six days. Only fourteen football league matches were postponed that Saturday, the best day for professional football in eleven weeks, but there was still no chance of play at the Shay, home of Halifax Town, just twelve miles from home. Instead they opened the ground as a public ice rink attracting hundreds of would-be Torvill and Deans. A few days later the coach of Halifax Town, Clive Baker, who was also a canoeing enthusiast, was able to paddle his canoe as floodwater now covered the pitch to a depth of two feet.

The night of 5th March was the first frost-free night across the whole of the UK since 22nd December and the thaw of incalculable tons of lying snow was well underway. It hadn't simply turned milder but really warm as a temperature of 17C in London on the afternoon of 6th March testifies. Together with some heavy rainfall across southern Scotland and the far north of England the consequences were inevitable as the River Kent in Kendal was suddenly transformed into a raging torrent. At Shrewsbury the Severn rose five feet between midnight and breakfast. A day later two hundred families were evacuated from their homes at Morpeth in Northumberland when the River Wansbeck burst its banks flooding the main streets. By the end of the day water supplies and electricity supplies had been cut-off.

Fears that the winter would end as dramatically as it did in 1947 with gales, torrential downpours and widespread flooding proved wide of the mark with only those incidents just mentioned of any significance.

There were many heartaches throughout the long winter of 1962-63, a winter so severe it is now being reported by the experts as a 1 in 1000 years or more event in these times of climate change. Prior to 1850, closer to the last mini ice age in Britain, such a severe winter would be classified as a 1 in 100-200 year event. The sporting calendar was in tatters for almost three months and understandably there were calls for the professional football season to close down in mid-season.

Horse racing was badly hit. Betting shops, legalised by the Betting and Gaming Act of 1961, found that they had nothing to bet on, excepting the occasional meeting in Ireland. Small, independent shops went out of business.

Courses affected by the bad weather had all been granted an extra fixture when conditions allowed but it was scant compensation. For the majority who hadn't taken steps to insure against abandonments it was a financial disaster. Owners still had to pay their training fees for horses that wouldn't be able to run and for the trainers themselves exercising the horses was a nightmare. Hard frosts or deep snow prevented the normal daily routine leaving the horses some way short of being 'match-fit' when racing was possible. For the jockeys it was even worse because if they didn't race they didn't get paid.

Julian Wilson, BBC Television Racing correspondent for thirty-two years until retirement in 1997, has been a long-time friend on and off the screen and I wondered what he could tell me about that racing crisis. He went on to explain that "no horseracing took place in England between 22nd December 1962 at Uttoxeter and 8th March 1963 at Newbury, ninety-four meetings being lost in total because of the ongoing snow and frost. Ayr, on the West Coast of Scotland, was able to race on 5th January and occasional meetings took place in Southern Ireland. The really severe weather set in on Boxing Day, frost causing the abandonment of the Kempton Park Christmas Meeting. Heavy snow arrived on December 27th with the north of England severely hit."

Bill Mather, who later trained in Yorkshire but at the time was a stud worker near Gisburn, Lancashire, recalls: "It was so cold that the soap froze to the washbasins, and the pipes in the cloakroom kept springing out of the basin connection. The snow had drifted and cut off the road from the village to the main stud. We had to cut across the fields to get to the village. I remember collecting the milk from the stud groom in a blizzard. This went on for over a month. Even the sea froze over on the east coast."

Julian mentioned to me that stables kept twin heat lamps on throughout the night. Nonetheless water buckets and bowls were frozen solid in the morning and pipes burst regularly. Trainers who could access a main road would transport horses to the nearest beach to exercise. Others rotovated the snow and laid down straw to create exercise areas:

"Tommy Robson, a leading trainer at the time, based at Penrith, drove his horses to Skinburness on the Solway Firth, to gallop. Noel Robinson, who trained near Berwick-on-Tweed, was a regular visitor to Holy Island Sands. Jockeys' incomes were severely hit, and some took jobs outside of racing. One or two travelled to Pau or Cagnes-sur-Mer, in the South of France to ply their trade."

For millions of horse racing enthusiasts the pinnacle of the National Hunt Season is not the Grand National but the Cheltenham Festival scheduled for March. Julian reminded me of what happened in 1963:

"The Cheltenham Festival began just four days after the resumption of racing. It was widely believed that the Irish would hold an immense advantage, having enjoyed spells of milder weather. The reverse was the case. English-trained horses won six of the seven races on Day One (Arkle was Ireland's only success); England (and Scotland) four out of six on Day Two; and five out of six, including the Gold Cup (Mill House) on the final day. Although the advent of race sponsorship, in the late 1950s, had created many new, televised, big-race opportunities throughout the spring, it was 1963 that underlined the wisdom of producing a 'fresh' horse at the Cheltenham Festival."

Away from the cauldron of top level sport, 'Wayward Wind' sung by Frank Ifield had been top of the UK charts for three weeks as winter changed to spring. There is no doubting the influence the wind direction had on determining the severity of the winter just gone. In meteorology there is an expression 'every wind has its own weather' but it seemed that the Siberian or Arctic winds dominated throughout those three harsh months.

The new number one on 12th March 1963 would be 'Summer Holiday' by Cliff Richard and the Shadows. A little premature perhaps but 52 million people now wanted to put the ravishes of winter behind them and think of what might lie ahead weatherwise.

# 3.

# A CAREER IN WEATHER AND MEDIA

It was a glorious summer's day in 1970 when I took the number 44 bus from Manchester's Piccadilly Gardens to Ringway Airport nine miles south of the city. Not only was 24th August my first day of employment as a scientific assistant but it was also my dad's 49th birthday. Fresh-faced, but with some apprehension, I arrived at the Met Office which was situated on the second floor of the control tower overlooking the apron between the domestic flights terminal and the international terminal. It was an exciting place to start my career but there would be many twists and turns ahead of me. Television was a long way from my mind.

I was introduced to the boss, George Band, who was sitting in the corner of his office peeling an orange. He was close to retirement but had always been an enthusiastic cross-country runner and as he stood up to shake hands I could see he was quite willowy and didn't carry an ounce of fat.

"Nice to meet you John, can you run fast?"

This was a pleasant enough introduction but not exactly the sort of question I was expecting, although it did cross my mind that if the forecaster had got it wrong that day he should be able to get away quickly from the angry pilots. I probably replied with something like "not really, I hated running at school". Seeing my puzzled face George immediately explained that he organised the annual staff marathon around the airport perimeter and he was sounding me out as a new recruit.

I was then invited to sign the Official Secrets Acts as I was now working as a Civil Servant within the Ministry of Defence. In the unlikely event that war broke out I would be employed as part of the RAF with an equivalent rank. Following a few more parish notices I was introduced to other members of staff on duty that morning and shown around the department by John Taylor, a scientific assistant for almost twenty years. What he didn't know about the job wasn't worth knowing, but I was first shown the kitchen where everyone shared duties at making the tea and coffee. Like many so-called 'outstations' this was a twenty-four hour operation and with the exception of a few in administration most of the staff were on shift-work. I would be no exception although initially I was a day worker.

The control tower had windows all round so the views of the distant Pennines were magnificent. Moreover, the second floor had a balcony on the outside so the weather observers and forecasters had 360-degree vision, absolutely vital when in constant touch with air traffic control above. My main duties in these early days would be learning how to observe the weather, sending half-hourly reports to the communications centre of the Met Office at Bracknell and to air traffic, plotting weather charts and copying charts for the forecasters to brief aircrew.

More menial duties would include making the tea and coffee, collecting the daily delivery of milk and maintaining the instruments down in the Stevenson Screen by the apron close to the aircraft. My first two weeks at Manchester Airport were blissful with perfect weather. If this was work then I couldn't get enough of it, but the shift-work was still to come.

On day work I was earning the princely weekly wage of little more than £10 but with weekend work and shift disturbance allowance in the near future that figure would escalate.

From 7th September, for the next four weeks, I was to attend the Met Office training school within a complex of government buildings at Stanmore, Middlesex.

Throughout much of the four weeks the weather was sunny, although I remember several mornings began foggy with a heavy dew. My digs had been booked just a mile away close to Edgware town centre so it was an easy walk up to the college – a walk I shared with two other members of the Met Office, Pete Davies and Jerry Capstick, who were staying in the same house. Pete was also at Manchester Airport but on a more senior course to me, and Jerry was an instructor at the college, co-incidentally a fellow Burnley fan from Settle in North Yorkshire. The house was owned by Mrs Russell – a keen West Ham United fan as I recall who was partial to a bottle of Guinness. Along the road was the pub which I was told was often frequented by Roger Moore of James Bond fame, although I never saw him myself.

My course gave me a thorough knowledge of the kind of work I would be doing when I returned to Manchester on 5th October but I had to pass the course as part of my probationary period. Apart from a short family holiday in Kensington two years earlier this was my first visit to London on my own but I was relishing the opportunity. I was now on the sharp learning curve of life as I desperately sought to leave behind the shy teenager from West Yorkshire and gain the confidence I would need for what lay ahead. At lunchtime many of us on the course would have a beer at the pub opposite the college and meet up with people from different government departments.

For just one week Lynn from Inland Revenue was the centre of my attention before she returned to Surrey never to be seen again.

As for the course it ended with a written assessment and an exam which I shared in top spot with Pam Street from London Weather Centre. New friends were made on that course who I would keep in contact with for many years to come as our paths crossed, either in person or speaking on the phone.

**Front row, second left, in this group photo taken during the Met Office training course at Stanmore in 1970.**

It was with renewed confidence I returned to Manchester Airport, initially doubled-up with more senior assistants and then eventually on my own. Regular contact with aviation personnel meant that I was now using the phonetic alphabet so I came to be known as 'Juliet Kilo' by air traffic control. As time went by I would meet my counterparts in ATC where 'Charlie Bravo' became my closest point of contact. Obviously she was not really a 'Charlie', much more of a Christine, and we met up outside work when shifts allowed.Unfortunately shift work impacts greatly on your social life and it also meant I had to live locally in Didsbury with the Parker family where I shared many good times for almost two years. I played cricket at every opportunity for the local club and returned home to Todmorden every ten days or so when the roster made it worthwhile.

When I left school with Physics and Applied Maths 'A' Levels it allowed me to begin my career in meteorology but I knew that further education would be required to progress to where I wanted to be as a forecaster. The truth was I was tired of being at school and needed to be more of a free spirit. For educational and social reasons I attended Stockport Technical College one evening a week where I eventually achieved my third 'A' Level in Pure Maths. To be fair to the Met Office they were very good at allowing staff to gain further qualifications by allowing special leave. I kept in close touch with my old friends from school, occasionally spending weekends in the Halls of Residence at Leicester University with my old pal Roger Slater and proving there was life away from shift work. There is no doubt it can be very tiring and I always had the desire to get back into full time education eventually, hopefully with the knowledge that I had a job to go back to.

A letter dated 9th August 1972 from personnel management, now human resources, at Headquarters gave me the opportunity to attend college full-time. I was granted special leave without pay for four years with effect from October, the third year being in industry.

Their letter stipulated that "Whilst on special leave I would remain a Civil Servant and stay on the strength of the Meteorological Office, but time so spent will not count for superannuation."

Bizarrely I was also issued with a posting notice indicating that I was due to return to Manchester Airport in August 1976. It seemed an eternity away and everyone knew I would never return, although I did make a few social visits during the holidays to meet my old friends and colleagues. One of the senior forecasters, Ron King, gave me a lift into Manchester one afternoon and he asked me with an air of resignation, "You're one of the best assistants we've had here for a while but why on earth did you want to join the Met Office?".

My reply was from the heart: "I am really interested in the weather and I want to be a forecaster like you one day, perhaps even working in other parts of the world".

There was a time when three year detachments were common, although it didn't always suit family life. For me that was a long way in the future but I remember being slightly disappointed that Ron should ask me the question, as if the Met Office wasn't what it used to be. Not the organisation he had been a part of since shortly after the war. Undeterred I knew it was the life for me.

So in early October 1972 I went back to being a student at Lanchester Polytechnic, now Coventry University, enjoying life to the full. Bands regularly played at the college on Saturday nights and on my very first day the mighty sound of 'Steppenwolf' and 'Born to be Wild' filled the air as they prepared for their gig in the evening. I suppose personally that song title did not really apply to me but I knew what I liked. I also had a little more cash in my pocket than most students as I was receiving monthly tax credits and because I was a couple of years older than most, and had 'been around a bit', I also became an 'agony uncle' on a few occasions. It was clear that if I hadn't gone back to college I would have missed out on some great experiences, my philosophy being 'work hard and play hard'. It might not be original but you do need to put the effort in so you can enjoy what student life has to offer and I wouldn't have missed it for the world.

The great benefit of already being employed, albeit without pay, was that after the first year I returned to the Met Office for a two months detachment at RAF Shawbury in Shropshire. This was a great opportunity to save a bit of money because I was now 'without wheels'. I had lost my first car, a Hillman Minx costing £80, in an accident near Rugby back in November but it had been invaluable during my time working unsociable hours at the airport.

Although the job was very different to before it was good to keep in touch and it was also a 'walk down memory lane' for my dad who was in the army during the war at nearby Doddington.

The pace of life during those two summer months in rural Shropshire was very steady. A few evenings in Shrewsbury and walks in the countryside were about as hectic as it got but the peace was disturbed one Saturday afternoon. Usually nothing much happened at the weekend but this day I witnessed for the first time the whole of the Red Arrows display team landing in front of me. To my amazement the leader, known as Red One, came over to the Met Office for a chat with the forecaster. His name I always remembered was Squadron Leader Dick and when my family and I were given a conducted tour at RAF Scampton, present home of the Arrows, in 2006 I discovered a photograph of the Team that year led by the same Squadron Leader. As I say nothing much happened during that two months detachment but at least my memory didn't fail me.

So in October it was back to Coventry for my second year, to my mind the best year as a student because everything is familiar and you know a lot of people by now and more about the area. So a brilliant year echoing to the sounds of Pink Floyd, Genesis, Barclay James Harvest and Chicken Shack with Mike Oldfield's Tubular Bells playing constantly in Tony Stead's room next door to me. In our own little enclave in the Halls of Residence our doors were permanently open and music filled the corridors. For some reason many of the physics students had all been grouped together although we were not all from the same year. Rob Smith, Pete Thomson, Steve Burge, Phil Tattersall, Mike King and many more but I remember them all as if it was only yesterday.

After a successful second year I wrote to the Met Office on 18th September 1974 with renewed confidence that I would eventually attain my degree. With this in mind I contacted personnel for consideration that my final year would be classified as Special Leave With Pay, which would obviously benefit me financially. References were obtained from my tutors and to my delight, after an interview at Bracknell in December, my request was granted.

In the meantime I was being paid for my third year placement at the Meteorological Research Flight based at the Royal Aircraft Establishment,

Farnborough. My work mostly involved working in the laboratory so I could use some of my physics knowledge. Many of the people there were involved in climate research with particular emphasis on atmospheric conditions within the tropics, notably in Senegal. For me that year was more of a means to an end but it brought my first contact with Farnham Cricket Club just a few miles away in Surrey.

My home town club at Todmorden was described by Johnny Wardle, the former England and Yorkshire cricketer, as having the most attractive cricket ground in the league. At Farnham I discovered another pretty ground situated by the castle on the edge of Farnham Park. It was idyllic but as it had a very pronounced slope away from the castle side of the ground I always made sure I was fielding on top of the hill!

At the end of the good summer of 1975 it was time for me to return to Coventry for my final year of study. In some ways it was a difficult year because many of the people I had known at the college had now left, although there were still plenty of civil engineers and scientists who had been on a similar placement as part of their course. Some of them I have kept in touch with to this day, which is a tribute to the great time I had there.

In my final year at college the great card game of life had dealt me one final card. I couldn't have predicted it at the time but this card held the power to completely change the content of my life. But it was a card that wasn't to come into play for another twelve years. More later.

Before my final exams I received a letter from Personnel dated 24th March 1976. The details of the letter were shattering. It read, "Due to a number of circumstances we are at present overmanned and there would be no promotion board this year."

This meant I would have to remain in the same grade, now known as assistant scientific officer, for a further twelve months even with a degree to my name. Obviously this news did nothing for my concentration or morale, but eventually on 18th June I completed my degree course and everyone said their goodbyes at Coventry, just as the hottest summer on record was beginning. For the next six weeks I soaked up the sun and did next to nothing living back at home, although by now the family had moved across the border close to Littleborough in Lancashire. I told them the weather would be better but I think it was because the council rates were lower! We have always 'watched our pennies'.

I returned to Farnborough on 2nd August with news of my modest Honours Degree but relieved all the same. What I had to do now was remain in the lowest grade, as a graduate, until a promotion board was arranged in 1977. This was easily the worst year of my working life, largely borne out of frustration that I couldn't get my promotion to then attend a forecasting course to pursue my dream.

**Todmorden Cricket Field in 1975 – in perfect weather.**

One thing I did manage was plenty of exercise. Along with four other work colleagues and lads living in the area we arranged a trip to the Yorkshire Dales in March that year and again in 1978 to attempt the Three Peaks. We hired an old farmhouse owned by Lancaster University where Mike Clement, one of our team, had studied. It was basic and damp but the oven worked and we could stay dry in sleeping bags. It was a fantastic experience and kept me close to the weather, which was always cold with short-lived blizzards thrown into the mix.

I also had more cricket matches for Farnham together with representative games for the Royal Aircraft Establishment as part of the Ministry of Defence cup competitions.

Eventually I was invited to a promotion board at 11.15 on 26th July 1977 followed by a letter confirming a successful interview on 9th August. I was now a Scientific Officer.

The next stage was another course at the Met Office College, no longer Stanmore but now a more rural setting at Shinfield Park, Reading. This would be the residential Applied Meteorology Course which would last eleven weeks from 19th September and covered my subject comprehensively. Following successful project work and a final exam there would be more frustration because at that time there was already a surplus of forecasters within the Met Office. So it would be a job in the fluid dynamics department

**Attempting the Three Peaks Walk – and getting to the summit of Whernside.**

at Bracknell until such time as a forecasting course was required. My job entailed "Running experiments using a thermally driven annulus to investigate the behaviour of baroclinic waves in a fluid".

You see the job was good, but it wasn't that good, although being Monday to Friday, 9 till 5, I could still play a lot of cricket. Now I had the opportunity to play for the Met Office as well as my regular 'cut and thrust' weekend matches for Farnham. I suppose by now I was part of the furniture at the club being a second team regular and occasional first team Sunday man. Often playing three times a week my form was as good as it ever got, culminating in the team winning the Surrey Cricketers League and in 1979 my taking seventy-six wickets, chipping-in with a few runs along the way when needed.

My final game for Farnham was away at Staines and Laleham, on 1st October 1979. It was so late in the season that dew was forming on the outfield by the start of the second innings and I had never played a game of cricket in October, before or since. I have no idea what the result was but we probably won; it was such a successful side, and it was the end of a very special era for me. Some good friends, great times and without question my most successful period as a regular league cricketer. When the captain threw me the ball I got on with it, often uphill into the wind. If I was required to stay down in Farnham and play again the following day I jumped at the chance – for some reason I often played even better.

Eventually I got the career break I wanted when invited to attend a Forecasting course at Shinfield Park which ran from 22nd November 1979 to 25th January 1980. We did have a break over Christmas and New Year! This course would have far-reaching consequences on how my career developed.

Two weeks before the end of the course we were given the choice as to which branch of forecasting we felt would be most suitable for a further six months 'on the job' training period. The choice was either public service forecasting at a weather centre or aviation forecasting at a military airfield or commercial airport. I really didn't have a strong opinion either way but the only way I could see myself getting onto the housing market was to move north. I had enjoyed my two years at Manchester Airport in the early seventies so I opted for a position there.

To my surprise I received an official posting notice from Personnel Management in which I was asked to report for duty on 28th January 1980 at Nottingham Weather Centre. I was happy enough with the location, after all it was well away from the most expensive area of the country, but a little baffled. Phil McGarry was also on my course and he had been appointed Manchester Airport despite his first choice being Nottingham Weather Centre. Notice a certain symmetry with the final outcome which we discussed in the college bar the evening prior to a visit from Personnel. At our personal interviews we could discuss our postings and see if there was any flexibility in the decisions.

Phil and I agreed to switch our postings if that was acceptable to Personnel. I was first in and thanked them for the opportunity to go to Nottingham but then explained that ideally I would like to swap with Phil McGarry. The bombshell from them was that I had been selected by my instructors as someone having 'media potential' and Nottingham needed an additional television presenter for evening inserts into 'BBC Midlands Today'.

The phrase 'bolt from the blue' hardly does justice to my reaction to the news. Maybe I was no longer the shy teenager everyone remembered during my life at school but I was still a reluctant 'entertainer'. Not for me the bright lights, although inwardly I had to admit that during some of the filmed briefings and presentations on the more recent courses my sense of humour had occasionally outweighed the shear trepidation I felt inside.

So it was a decision that was irreversible. I would go to Nottingham and have a television audition as soon as possible. For the first month or so I stayed with Jamie Thomas in Leicester who had also graduated at Coventry as a Civil Engineer back in 1976. Then I moved into a bed-sit at Lenton close to Nottingham city centre during what was one of the warmest and sunniest springs on record in the East Midlands.

More important was the forecaster training I was now enjoying and also the imminent audition which I was more apprehensive about. It was

**The forecasting course at Shinfield Park in the winter of 1979/80 had a profound effect on how my career developed. I'm third left, back row.**

probably March when I travelled the six miles down to the BBC studios in Nottingham with another candidate, Graham Butler, together with the Head of the weather centre, Ernie Pepperdine. We would be joined by David Stevens who was Head of presentation at Pebble Mill in Birmingham. David was a thorough professional who immediately settled our nerves by suggesting we go for lunch in the bar at the Nottingham Playhouse just round the corner from the studio. We enjoyed a buffet salad and a couple of beers before returning to the job in hand, namely the BBC having to choose between Graham and this reluctant broadcaster.

We were given several opportunities to impress under the studio lights, not as weather forecasters but simply as potential presenters. David asked us to explain what we did at the weekend or talk about our hobbies and interests.

Could we cope with the timing constraints, maybe anything from thirty seconds to two minutes? Could we talk without hesitation, repetition or deviation? Above all did we look comfortable under pressure in a strange environment?

We reviewed the tape with additional comments from our boss and also Terry Bracey, the technical engineer driving the desk. From my point of view I was pleasantly surprised with my efforts and when Terry asked me privately what I thought he graciously agreed that the position should go

**Early TV days in Nottingham, 1980.**

to me. "I was more natural," he said.

On the day of my first appearance I was accompanied by another forecaster, Len Moloney, who had already done some television in recent months and would do so again if I lost my nerve. I didn't realise it at the time but I was actually being trained to replace Len in the near future when he moved elsewhere. At the Nottingham studio the Head of Regional Television had also arrived from Pebble Mill. His name was Michael Hancock.

I felt there was a lot of pressure on me to do well and Michael offered me a whisky (or water) to steady my nerves a few minutes before the programme went on air. Probably to his relief I opted for the water but as I chatted in the studio with him I noticed one thing. My heart was racing inside but my glass of water was perfectly still in my hand. At that point I knew I could do it and I just wanted to get 'on air'.

Tom Coyne was the main presenter of 'Midlands Today' at the time and David Stevens the newsreader. Prior to my weather forecast was the sports bulletin which was highlighting Nottingham Forest's European Cup Final against Hamburg in Madrid later that evening. The cue to me from Tom would make mention of the match.

It couldn't have been better for a debut because I could settle my nerves with talk of the weather in Madrid being very warm, which it was, and also predict that Forest would win the game, which they did 1-0. My thoughts about the weather in the Midlands were also included but that didn't seem to matter quite so much, although I must have got the message across. I had made my mark; I would be remembered for getting the score right and also predicting a fine evening in Madrid in late May! Michael Hancock was happy and so was I.

As the months went by I gained experience in forecasting and shared television duties with some of my colleagues at the weather centre. Of course these were the days of magnetic symbols and basic two-dimensional charts, usually one for tonight and one for tomorrow with weather symbols arranged across a map of the Midlands. It was impossible to be ambitious with the graphics but at least you could show-off your skills as a presenter by making the presentation interesting and topical with an awareness of what was happening around the region. Occasionally we left the confines of the studio and did an outside broadcast, the annual Goose Fair at Nottingham in early October being a particular favourite of mine.

On 13th October I eventually received confirmation that I would be staying at Nottingham after successfully completing my further training. About the same time, after months of negotiation within the Civil Service, a revised structure for television allowances was agreed which meant that every day I appeared in front of the camera additional money could be claimed. Modest by television standards but a useful monetary incentive, it helped secure my first house in early 1981, a modern semi just over five miles from work.

Radio broadcasts were an additional requirement as the Nottingham Weather Centre was responsible for all media output in the East Midlands at the time. Radio Leicester, Radio Nottingham and Radio Derby were already established with Radio Lincolnshire the latest acquisition and Radio Northampton coming later.

Although my appearances on the cricket field were now a victim of shift work it was a great honour to join forces with the Radio Derby cricket team who played several twenty overs a side evening games through the summer and occasional Sunday afternoon matches. Their captain was Ashley Franklin, an experienced presenter with a keen sense of humour who would read out his match report on the radio the following morning. Self-appointed wicketkeeper (no-one else would do it) was the humourist Deric Longden who, in his own words, was "almost educated" in Derbyshire but had moved south from Chesterfield to Matlock. Tragedy struck the family when his lovely wife Diana was diagnosed with ME and confined to a wheelchair for the last ten years of her life. She came along to some of the matches but life was becoming increasingly difficult and she eventually passed away in 1985. Deric wrote *Diana's Story*, which eventually became a screenplay on the BBC in 1993 with the brilliant Julie Walters playing the part of Diana in 'Wide-Eyed and Legless'. The sense of humour still shone through even in tragedy.

Deric's gentle wit was eagerly anticipated on Radio Derby and his weekly reflections on the latest Matlock Town performance was a joy to hear.

In February 1983 it was back to the Met Office College, this time for my Advanced Forecasting Course ending on 25th March. Further on-the-job training was then required for four months at another location, which in my case would be a military establishment.

RAF Linton on Ouse near York was my appointed detachment, arriving immediately after the Easter holiday. As a Scientific Officer I was allowed to stay in the Officers Mess with all the benefits attached, namely a cheap bar, good food and plenty of it. I also kept my bike in my substantial ground-floor room, which gave me ample opportunity to sample the delights of the Yorkshire countryside and an occasional six-mile ride into York. It was a marvellous four months, despite poor weather, with golf and horse racing enjoyed to the full in my spare time. I enjoyed plenty of banter with the RAF as well and even managed a flight in a jet provost on 21st July, aircraft number '61' XW 405 flown by Nick, my very able and affable Squadron Leader. He described our one hour trip as "the weeniest sortie ever" but I was never very good at heights, especially upside down in a spiral!

On my return to Nottingham I was promoted to Higher Scientific Officer but there was to be a bitter blow 'waiting round the corner'. For almost four months I had briefed aircrew and presented weather information to an audience of RAF personnel, including the Wing Commander who was equally concerned about the weather for his round of golf on Sunday morning. But the media work was absent and that was where I really saw my future in the Met Office.

By early August the news came through from the BBC at Pebble Mill that they were dropping the live weather presentations for scripted graphics. Local radio broadcasts would still survive but no television. For many months I missed the twice-weekly trip into Nottingham for the early evening broadcast and life was not as fulfilling as it had been.

In the meantime my name had been noted and placed on a list of regional weather presenters who could possibly graduate to national exposure in the future. I was invited to the BBC at Television Centre to meet the existing team led by Jack Scott. We had lunch together in the infamous BBC canteen followed by an informal interview and a brief audition in the afternoon. It was a privilege to be invited but at that stage I had no interest in moving to London. My heart wasn't in it and frankly I probably didn't make much of an impact. But Jack always saw the potential and had been impressed by some of my terminology on presentation tapes he had reviewed earlier. Maybe I let him down at the time but I remember enjoying the train journey north again!

Luckily, Central TV were in the process of building new studios just to the south of the city, ironically within one mile of my original bed-sit at Lenton. They were planning to cover the East Midlands with their innovative news programme in direct competition with 'Midlands Today'. The Met Office was approached by Central TV for weather presenters. Auditions were arranged which at this stage were in their temporary makeshift studio at Giltbrook, very close to the Weather Centre. Of course they had a good idea who was available and what we could do but eventually a decision was made.

Jim Gould, John Kettley and Dave Squires was the chosen team of weather presenters for Central TV in the East Midlands, from August 1984. A media launch was arranged by the excellent Press Office at Central TV and the new studio complex at Lenton Lane was now in full production. Our remit was to cover the weather for the whole of the East Midlands leaving others, including Ian McCaskill for a while, to concentrate on the West Midlands from Central's other studio in Birmingham.

With the introduction of improved computer graphics I was now leaving behind magnetic symbols so revered at the BBC and instead making use of a system known as chroma key at ITV. This required the weather presenter to stand in the studio in front of a large screen with small television monitors at either side. Although the presenter couldn't see anything behind him the side monitors allowed a view of the transmitted picture and where to point. The tricky bit was the 'eyeline', which had to be just right and also to wear clothes that were not the same colour as the screen otherwise the graphics would appear where your jacket or tie should have been!

Television was already moving at quite a pace and for me this was another fresh challenge but a very exciting one. Instead of presenting the weather from a remote studio I was now more integrated allowing more banter with the other presenters and more flexibility within the programme.

Friday evening was a favourite because I was allowed more time with the weekend ahead and Dave Bartram of 'Showaddywaddy' fame often presented an entertainment item just before the weather so we struck up a good rapport. Jimmy Greaves, Derek Thomson and Tony Francis were also regular sports contributors although Andy Craig was usually the main presenter.

The large studios within the complex brought several high-profile shows to Nottingham, which made the place a more exciting environment in which to work. 'Blockbusters' with Bob Holness, 'The Price is Right' with Leslie Crowther and 'Auf Wiedersehen Pet' were just a few of the iconic programmes being made at that time and it was quite normal to catch celebrities in make-up or the restaurant. Tickets were freely available to view the shows so the opportunity to broadcast with ITV at Central Television afforded a refreshing departure.

**Keeping pace with new technology and using side monitors to indicate where to point. It had its tricky moments.**

The following March I received a letter from Roger Hunt, Head of London Weather Centre, inviting me for a further audition at Television Centre. It was 16th April 1985 when about fifteen candidates were introduced to Bill Giles, the successor to Jack Scott who had now reached retirement age. The day was to be a pleasurable experience with wine flowing and food to eat but unlike last time my inner ego was saying that now could be the right time to think about national television. After all I was still enjoying the experience and this was a time when weather presenters did get the chance to appear on other programmes, which would be fun if asked.

By accident or design I was the last of the candidates to audition that day and it was now late in the afternoon. It may not have been a coincidence that when my turn came there was a problem with the stopwatch. I knew my allotted time was over-running but I kept going; that is what you have to do in 'live' television until the director says otherwise. Eventually I was told to 'wind-up' after almost three minutes but it all happened naturally.

My future appeared to be sealed and a further letter dated 24th June from Personnel confirmed that I would be leaving Nottingham and moving to the London Weather Centre. A new chapter was about to be written.

# BURNLEY FOOTBALL CLUB – A LOVE AFFAIR

Burnley Football Club joined the newly formed Football League in 1888. The original gang of twelve also included Notts County, Derby County, Aston Villa, Wolverhampton Wanderers, West Bromwich Albion, Stoke City, Everton, Accrington, Preston North End, Blackburn Rovers and Bolton Wanderers.

It is still clear in my mind the very first occasion I began looking at the football results in the newspapers; after all the amount of sports coverage in the distant days of black and white television was minimal compared to today's terrestrial and satellite coverage, not to mention internet.

On the Sunday morning our local newsagent, Ronnie Walton, would do his rounds with an enormous sack of newspapers wrapped around his shoulders. As usual he was making his way along Commercial Street, only about 300 yards from his starting point on Halifax Road, when I intercepted our regular order, *The People*. No doubt he was quite grateful that his sack was already feeling more manageable before he headed back to the shop for the next lot. Checking on Saturday's results I stood open mouthed as I read aloud to my pals – Burnley 8 Nottingham Forest 0 – a throwback to those Victorian days in the Club's first season. The date was 21st November 1959 and I knew immediately this was a team I wanted to see play. Burnley was only nine miles away so why shouldn't my dad take me along at the earliest opportunity. Actually both my parents had supported the team, and even walked from Todmorden to Burnley in their early days of married life shortly after the war. It was Good Friday 15th April 1960 when dad eventually took me to a live match against Leicester City at Turf Moor, home ground of Burnley FC.

It was a balmy spring day as we stood in the family enclosure in the main stand – just amazing that I should remember what the weather was like unless I was looking through rose-tinted glasses!I was still only seven years old and barely able to see over the perimeter wall bordering the pitch but I found the whole atmosphere electric – the smell of liniment and the sound of hits from the charts blasting across the ground on the loudspeakers. Top of the pops for four weeks through April that year was Lonnie Donegan with 'My Old Man's a Dustman' – thank goodness we were about to see a renaissance in popular music with the Mersey Sound just a short time away! There was to be no repeat of the Nottingham Forest goal avalanche back in

**Heady days in 1960. The final match of the season saw Burnley travel to Maine Road, Manchester, to win the League Championship 2–1. The club's winger Meredith scores the vital second goal. (PA Archive)**

November but Burnley did win the match 1-0, a goal scored just in front of where we stood by right winger John Connelly, an England International as well as Burnley hero. More significantly there were just a handful of games left in the 1959-60 season and it was becoming increasingly likely that this small mill town from industrial Lancashire, population just 80,000, was likely to have a big say in who would be the Division One Champions.

These were heady days and the final match of the season saw Burnley travel the short distance to Maine Road, home of Manchester City. It was a midweek encounter and out of bounds for a young schoolboy but like thousands of other passionate supporters I followed the unfolding drama on BBC television.'Sportsview' hosted by Peter Dimmock was the flagship sports programme in those days along with 'Grandstand' on a Saturday.

Midweek games always kicked-off at 7.30 in the evening and it was a game Burnley had to win as they started the game in third position. Wolves and Tottenham remained major obstacles just above them in the table but had already completed their fixtures. For those two clubs and their supporters it was just as nerve racking but they had been here before. They were both big clubs with huge reputations – for Burnley they were ninety minutes away from their first Championship since 1921. On 2nd May 66,000 fans packed into Maine Road with many more locked outside. Understandably the match was a tense affair although Burnley went into an early lead. The score stood at 2-1 to Burnley after thirty-one minutes leaving nerves to jangle for almost an hour before the final whistle blew. The result was quickly announced on 'Sportsview' with the resulting caption that Burnley were indeed the English Champions.Unfashionable perhaps compared with the city clubs like Arsenal, Manchester United, Chelsea and Everton – but this was my team!

Now at this point I need to explain something. Many people over the years have been puzzled to understand why a proud Yorkshireman with his roots in Calderdale has for most of his life been so passionate about a football team from Lancashire. To understand the reason you need to remember that my home town of Todmorden has always been referred to as the border town, some believing it should be in Yorkshire and others in Lancashire. After all it was for hundreds of years bisected by the River Calder and the magnificent Town Hall was actually built above the river in 1888. Administratively the town retained a split-personality with the battlelines drawn literally at the Town Hall and to this day it remains something of an enigma.When you send a letter to anyone living in the town the postal address is Lancashire; the cricket team is represented in the Lancashire League but with a white rose and a red rose as its emblem. But the district council is now run from Halifax, Todmorden being the far western outpost of Calderdale.

The fact remains that Burnley is just a few miles across the border from Yorkshire, probably four at the nearest point, and close enough not to be a problem with a team as talented as it was in my early days!

It was not until 11th February 1961 that I returned to Turf Moor for my next match, an epic confrontation against Sheffield Wednesday in monsoon conditions. The Division One league table that morning showed Burnley again well placed in fourth position with Wednesday just one place above, the clear leaders being Tottenham Hotspur. It was such an anticipated encounter that a big crowd battled the elements but the weather was so wet that the family enclosure was sparsely populated as thousands of fans tried to get under cover – not an easy job in those days with the Bee Hole end of the ground also completely open. The rain was torrential before the match leaving standing water all over the pitch, the groundstaff doing their best with garden forks to get the water to soak through. It was Burnley's first home match for three weeks, partly due to bad weather, so every effort was made to get the game on. These days the match would have been postponed without a doubt but in the sixties everyone was accustomed to playing even professional games on a pitch with green oases towards the corner flags but a broad strip of mud down the middle.  It often made for exciting games but it was virtually impossible to recognise most of the players as they trudged off after ninety minutes, and remember these were the days prior to substitutes.

When the match got under way at 3pm it soon looked grim for the Clarets as they quickly went 0-2 down – now I hoped the referee would blow his whistle and admit the conditions were just too bad – sadly he persevered. Captain Jimmy Adamson scored from the penalty spot after 23minutes with an equalising goal from John Connelly on the half hour.  Unfortunately there was further drama just two minutes later with another goal for the visitors and the teams went in at half time with Sheffield Wednesday leading 3-2. Out came the brass band at half-time as the rain relented but the second half continued with one more goal for each side. A heavy mist hung over the surrounding hills and the weight of the occasion was too much for Burnley to come back.

Thoroughly depressed, dripping wet and with the steamed-up bus windows weeping in sympathy, the nine mile journey across the border on an ageing double-decker personified an unwelcome defeat.

Nonetheless the 1960-61 season was another success for the Clarets, eventually finishing in fourth position in Division One, semi-finalists in both the League Cup and FA Cup with the bonus of European football as reigning champions.

Politically there were significant changes happening behind the scenes in English football when Jimmy Hill, chairman of the PFA, negotiated the abolition of the maximum wage, then £20 a week. Fortunately with average home crowds of 24,000 the club was in good shape to pay their players the increased wages and another fine season beckoned in 1961-62.

These were unquestionably heady days.  I continually pestered my dad to take me to every home match in the early part of the season because this was an adrenalin rush I couldn't live without.  The routine was the same for all Saturday games, namely pre-book the tickets for the football special leaving Todmorden outside Calder College at 1.45, returning from the main car park behind Brunshaw Road around 5.10. Never completely at home in the kitchen he also cooked a simple lunch consisting of 'chips, peas and something else' whilst mum went on her customary weekly shopping spree to Manchester or Halifax with my younger sister Susan.

It was yet another satisfactory start to the season but on 23rd September there were unsavoury incidents during the match against Everton.  Crowd trouble resulted in headline news in the Sunday newspapers as sixteen visiting spectators were arrested after bottles had been thrown onto the pitch.  Remember these were still the days of flat caps and rattles as thousands of football followers around the country made their fortnightly pilgrimage to see their teams play a home match.  The trouble that day was as rare as it was frightening for a nine year old, and reluctantly we didn't get to Turf Moor again until the Nottingham Forest victory by 4-0 on 31st

March. Further good results in league and cup matches brought more rewards for this excellent team, eventually grasping defeat from the jaws of victory and finishing runners-up in the Championship but more significantly a place in the FA Cup final in early May against their old rivals Spurs. This was Burnley's first Cup Final since they lost 0-1 to Charlton Athletic in 1947, their only cup win being way back in 1914 when they defeated Liverpool by the same score.

The team more or less picked itself consisting of household names now familiar throughout Britain: Blacklaw, Angus, Elder, Adamson, Cummings, Miller, Connelly, McIlroy, Pointer, Robson and Harris.

The Cup Final was always played on the first Saturday in May and the weather was invariably warm and sunny. It was in 1962 but sadly Burnley never recovered from an early goal by Spurs and the game was lost 1-3. Inexplicably I didn't even watch the match on television and to this day I really don't understand why – perhaps inside I must have felt it wasn't going to be.What I did instead was walk up the road to Todmorden cricket ground and spent five hours in pleasant weather watching a Lancashire League match, and now I couldn't even tell you who they were playing. But it was live sport, and maybe if I'm being honest cricket really was my first love with football just behind.

By the following season there was a changing climate in my world, and not just meteorological, despite the infamous 1962-63 winter being just round the corner. Bob Lord was the larger than life chairman of Burnley FC, local butcher and businessman. Obviously I never met him but those involved with the club knew who was the boss. Admired and yet feared by many, perhaps his surviving legacy was his decision to sell arguably his best player in those halcyon days of the late 1950s and early 1960s. To the thousands of Burnley supporters Jimmy McIlroy was God. He was an experienced Northern Ireland international and playmaker for this great team which had taken the football world by storm for the past four years. Despite average gates of 27,000 in the previous season and an excellent and respected youth policy, the cost of running the club was increasing as the effects of abolishing the maximum wage were being felt. On 5th March 1963 Jimmy McIlroy was sold to Stoke City for £25,000 and with him went thousands of loyal supporters who they believed had witnessed the end of an era.

The rebuilding of a new team continued over the following two seasons as several of the household names from the glory years were transferred to other clubs, or retired. I too must have sensed the 'wind of change' because I did not return until the local derby match against Blackburn Rovers on 9th October 1965. Burnley were top of the league coming into the match but it was to be a dreadful 1-4 defeat to their arch rivals by a quarter to five. Of course I was now old enough to go on the football special with my mates from the street or from school, no longer 'cool' as a young teenager to go with your dad!

Undeterred by the heavy defeat I was a regular at every home match and some away games for the rest of that season, even taking in a few reserve games along the way. Burnley finished the season in a respectable third place with a potent strike force of Willie Irvine and Andy Lochhead. For the next five seasons I would be going to Turf Moor as a season ticket holder.

I had the football 'bug' – of that there was no doubt – taking in away matches at Manchester United, Leeds United, Blackpool, Sheffield United, Manchester City... the list was endless.

There were many good performances but a few low points along the way and it was following a sequence of poor results in October 1968 that a new generation of young players was introduced to the first team. Back in the spring Burnley had become Youth Cup winners beating Coventry, their goalkeeper that evening being the legendary David Ike who may have likened himself to God but couldn't prevent two goals entering his net! Now blossoming stars like Dave Thomas and Steve Kindon were thrown in and the team won eight in a row, including a 5-1 thrashing of the seemingly invincible Leeds United. Ask anyone at the time what they thought of Don Revie's side and most football supporters would say 'arrogant' but 'ruthless' – most people were envious of this well-drilled, almost mechanical football team. Nonetheless it was a Burnley performance which to this day I believe was the best I ever witnessed, and over 26,000 packed the ground to see it.

The team was also just one step away from Wembley as they were drawn against Third Division Swindon Town in the semi-final of the League Cup, and thereby hangs a tale!

Remember Swindon Town had been a Third Division side for most of their history and Burnley had now been in the top echelon for twenty-one consecutive years. After the home and away legs the scores were level at 3-3 so a replay was needed on neutral territory, The Hawthorns, home of West Bromwich Albion in the Midlands.

It was shortly before Christmas, 18th December 1968, and the game was scheduled for midweek. Many of my mates at school were also keen Burnley fans and as a group we contemplated the idea of hiring a small coach and driver to travel the hundred miles or so to the match, after all it was just one step away from the final at Wembley and possible glory in the League Cup. Just one problem – we needed to leave school early that afternoon to get to the match with time to spare. So we appealed to the headmaster's better nature. Of course we were lower sixth formers and we already knew everything about life in general. Surely we could be trusted, along with our peers in the upper sixth, to get to West Bromwich and back in one evening and be back in school the following morning on time! Our trump card was that Bill Hall, one of our group, just happened to be the son of the headmaster so whether or not permission was granted we were all in rebellious mood and off we went for the replay.

It was a cracking match with a great atmosphere, somehow evening matches under bright floodlights in a steady drizzle always add to the ambience. The game went into extra-time so it would be even later when we got back home that night.  More important was the result, Burnley 2 Swindon Town 3, so not for the first time this Third Division outfit had jinxed my heroes and we all felt devastated.

It felt a long journey home that night and I suppose we were all half-asleep when we turned up for assembly the following morning, shaken if not stirred.

The final game of the season ended in a 1-2 defeat to Sunderland, a predictable outcome for a game originally snowed off after thirty minutes during the winter when we were leading 1-0 as the blizzard arrived.That was easily the shortest match I have ever seen!

For the third time in a row Burnley finished in fourteenth position at the end of the season and despite the introduction of several youngsters and transferred players this was another mediocre performance.  They were conceding far too many goals for sure.

The 1969-70 season started badly, bottom of the table at the end of September, then some decent results in the coming months saw them ten points clear of relegation by mid-February but there was about to be a major shock move.

Jimmy Adamson was the captain of the team in the glory years a decade earlier when Harry Potts was the highly respected manager and also a former player.  In February 1970 it was decided to introduce Adamson as manager, Harry Potts being given the new role 'upstairs' as general manager.

Amazingly the end of the season saw Burnley finish yet again in fourteenth position but Adamson was about to make a statement that would come back to haunt him for years to come: "Burnley will be the team of the seventies.  We are building one of the finest stadiums in the country and we have a great young team to go with it.  In the next few years we will win the Championship not once but several times."

This was a statement to gladden the hearts of all supporters but it did sound too optimistic. Not only had gates in recent seasons dropped alarmingly but the abolition of the minimum wage was now biting hard for the many small town teams. The rich were getting richer, the rest were having to sell players to get by, and Burnley were no longer part of the elite.

The writing was on the wall when Liverpool won 2-1 at Turf Moor on that first afternoon of the new season! I was due to start my career in the Met Office at Manchester Airport nine days later on 24th August so with shift-work on the agenda I didn't buy a membership ticket that season. Still I did get to a number of games, although sadly it was to be a season of heartache, despair and for the manager – regrets. Relegation followed after twenty-four years in the top flight of English football but for Bob Lord every cloud had a silver lining when he sold England International Ralph Coates to Tottenham for £190,000 – a deal he described as "one of the finest pieces of business enterprise ever pulled off by a football club".

Yes, Mr Lord was forever the businessman!

The club embarked on their first season for a quarter of a century in Division 2 with mixed emotions but there was a determination to return as soon as possible with firm financial incentives for the players. In the end they could finish only seventh and the performances throughout the season gave no guarantee that a return to the First Division was imminent.

However, with fewer injuries and more consistency from players like Martin Dobson and Leighton James the following season proved to be very successful and Burnley were crowned Champions of the Second Division at the end of April 1973. The team was beaten only four times – at Queens Park Rangers, Nottingham Forest, Sheffield Wednesday and at home to Orient but amazingly I was present to see three of those four defeats! By this time I was at Lanchester Polytechnic, now Coventry University, studying for my physics degree but I travelled to see the team play whenever it was practical.

It was a fantastic season in many ways. Even the defeat at the City Ground where Forest were convincing winners 3-0 was easy to take as I had also backed the winner of the Grand National that afternoon, Red Rum obliging at 9/1.

So a return to top-flight football in 1973-74 was a morale boost to my home area in the Pennines as it coincided with the enforced power cuts due to the miners' strike. Speed restrictions also meant that cars could not exceed 50mph even on the motorways, although my first car, a Hillman Minx, was already at the scrapyard having been written off in a minor collision months earlier. I was now reliant on coach and train.

On the field the season went well although it was sickening to see some of the treatment dealt out by some of the opposition on our potential 'match winners'. I remember being at Turf Moor on Tuesday 11th September (score ended 2-2) when striker Frank Casper was hacked down by the bully boys from Tottenham's midfield, an injury from which he never fully recovered and prematurely ended his playing career. Best result of the season was probably the 4-1 defeat of Champions Elect Leeds United in March but consistency throughout the campaign led to a creditable sixth finish, a performance made even sweeter by the subsequent relegation of Manchester United!

There was also a memorable run in the FA Cup and once again I made the journey from Coventry to Sheffield for a semi-final encounter with Newcastle United at Hillsborough.

A crowd of over 55,000 packed the stadium and I was surrounded by more people wearing the black and white scarves of United than I felt comfortable with. Remember these were the days of fairly widespread crowd violence in and around football grounds, perhaps unsurprising given the state of the country at the time with most people working a three-day week. In the end Burnley lost 0-2, probably a fair result although the game never really came to life and it was the sharpness of Malcolm Macdonald in front of goal that made all the difference.

The balance sheet at the start of the 1974-75 season showed a loss of £224,000 so Martin Dobson was quickly sold to Everton for £300,000, no doubt through necessity to pay for the new stand on Brunshaw Road and the increased wages.

On the field the team promised plenty after getting off to a great start but there was to be a sting in the tail. The greatest embarrassment occurred in the third round of the FA Cup on the 4th January when I again jinxed the team. Staying up north after the Christmas break it would just be a formality to see-off the 'crazy gang' from Wimbledon, non-League at the time. In the event almost 20,000 witnessed a major upset as Burnley went down 0-1, the local press suggesting it was "arguably the most humiliating result in the history of Burnley FC".

In retrospect this was the pivotal point of the season and arguably in the fortunes of the club for many years to come – and sadly I had been there to see it! The team finished the season in tenth position but worse was to follow in 1976.

That said it was a decent start to the new season and the game against Norwich on 13th September was one of the most entertaining I have ever seen. The game ended 4-4, a nightmare performance by both defences but for Peter Noble, one of the hardest working Clarets players there ever was, a personal triumph having scored all four. By the New Year the situation had gone from bad to worse with Leighton James transferred to Derby County and manager Jimmy Adamson (the team of the seventies claim ringing in his ears) resigning.

Weather-wise the summers of 1975 and 1976 were the hottest and driest for years leaving serious droughts conditions across much of Britain. Mirroring that situation, and following the goal fest against Norwich City early in the season, there was a drought in front of goal by the start of March. In the final nine games the team scored only six times and the inevitable result was relegation back to Division Two. Many people said there would be no way back, but recent triumphs in 2009 proved otherwise!

By now I was living in the south of England and it became less likely that I would be visiting 'The Turf' to see a game. Mentally the torment of events on and off the field in the past two seasons were beginning to curb my enthusiasm and from afar I could only see turmoil and more troubled times ahead.The next ten years were to be at best indifferent and at worst thoroughly depressing after the vintage performances I witnessed as a boy in the early sixties.

I could write the highlights in the period from 1976-86 on the back of the proverbial envelope; surprising really as there were some fine players at the club, notably Trevor Steven, Alan Stevenson and Billy Hamilton, but somehow the team rarely seemed to operate as a winning unit for more than a few games at a time.

As the hot weather continued through August 1976 a goalless draw in the first match at Wolves saw the eleven Burnley players lose 60lbs between them! But metaphorically it would get even hotter as the season went on for the Board and for the players as changes in personnel continued. After languishing near the bottom for some time they finished the season in sixteenth position followed by eleventh in 1978 and thirteenth in 1979.

There was no such thing as the 'red button' or text messaging in those days so however despondent I might have been as a supporter in exile it was often a case of waiting until 16.45 on a Saturday afternoon to get the final score – win, draw or lose.

It may be a cliché to say that Burnley was the first result I looked for but believe me it is true. Between 1980 and 1985 I had moved north again to Nottingham where I began my forecasting and broadcasting career so there was more opportunity to see the team play. But where was the incentive because on 3rd May 1980 a crushing defeat at Watford sent Burnley into the Third Division for the first time in their history.

So the 1980-81 season was a time to regroup but with average home attendances close to 6,000 the atmosphere in the ground was quickly disappearing, a finishing position of eighth being the springboard for the following season. It was sad to see opposition like Runcorn and Altrincham playing at Burnley because of the earlier entry in the FA Cup competition for Third Division teams, but we couldn't beat non-League Wimbledon all those years ago so they both deserved their chance.It was a season of success in the end culminating in a triumphant evening in front of 18,000 fans when the Championship was clinched against Chesterfield.

So 1982-83 promised plenty but it was to be relegation again at the first attempt. It was however a season of enigmatic performances because as the team struggled in the league it was successful in the cup competitions. The FA Cup saw progress to the sixth round before losing to Sheffield Wednesday but in the League (Milk) Cup there were victories against Bury, Coventry, Middlesbrough, Birmingham and Tottenham before losing in the semi-final to Liverpool. It was to be another twenty-five years before Burnley would repeat those kind of performances in the League Cup, in January 2009 losing in the semi-final to Tottenham this time.

Further relegation followed into the Fourth Division in 1985 and the agony for me and thousands of fellow supporters was complete. We would now be playing teams like Halifax Town and Rochdale, teams I used to watch 'for a laugh' on a Friday night or on Saturday when Burnley were playing away from home.

Could it possibly get any worse? Well it very nearly did on 9th May 1987, a date which could have brought the end to Burnley Football Club as we knew it. The average attendance for home matches that season was under 3,000, but for the final game against Orient a crowd of almost 16,000 came along to witness history in the making. After the worst season in the club's history we had to win against Orient but also their two main rivals at the bottom had to drop points.

So where was I on this final game of the season? Would I be at Turf Moor lending my support? Well sadly I was working at the London Weather Centre in High Holborn, two hundred miles away in distance but there in spirit as I kicked every ball that afternoon listening to the BBC Radio 2 commentary from the ground. For ninety minutes my thoughts were a million miles away from deepening areas of low pressure out in the Atlantic. This was a high-pressure afternoon, a pressure cooker of emotion and a fear of oblivion.

Crowd congestion and a fifteen-minute delay to the start of the match further fuelled the torture that was happening in East Lancashire. Having stood on the terraces hundreds of times since my first visit as a seven year old I could readily translate the audio commentary into a visual image as I listened alone at the weather centre. Every shot, every tackle and every corner were in HD, though of course we had to wait another twenty years for the real thing.

At the end of the first half Grewcock gave Burnley the lead and with more good news from around the country at half time there was a spring in the step as the players began the second half and I again listened to the commentary, having given myself a ten minute break to check on the weather situation.

Within minutes Britton made it 2-0 but then it was back at 2-1 with thirty-four minutes still to play. For many thousands of supporters like me this would be the longest thirty-four minutes of football ever witnessed in over one hundred years of club history. Shortly before 5.00pm the ordeal was over and relegation had been avoided, many people on the ground apparently in tears of joy. I felt the drama and the relief all those miles away as I punched the air, but this situation could never be allowed to happen again.

So that day in 1987 Lincoln City went through the football trap door, although they had remained above the bottom position throughout the season. Ironically we now live just outside the city of Lincoln having moved north from Hertfordshire in late 2005.

Alistair Campbell also has distant memories of the glory days when we were both young boys and even more than me he has maintained his commitment to the club. He believes the passion in the town was reborn after the great escape against Orient, the community realising the significance of what they nearly lost, maybe forever.

I had been a regular on national radio and television for around two years and rarely did I miss the opportunity on a relevant broadcast, for instance during Grandstand on a Saturday lunchtime, to give a mention of the weather for today's match at Burnley. These constant references did not escape the great and the good at the club itself and it was a privilege beyond my wildest dreams to be invited to a home game in the 1988-89 season. Along with my wife Lynn we gratefully accepted the invitation to take a seat with the Directors for the home match against Darlington at the end of a week of fell walking in the Lake District.

15th April 1989 would turn out to be a memorable date for all the wrong reasons. For the record the weather was pleasant enough but only 5,500 turned up to watch a desperately poor performance and a 0-1 defeat to bottom club Darlington. There was little to cheer on the field but I thanked the chairman Frank Teasdale (Bob Lord having died in late 1982) for inviting us.

"We've enjoyed it," I said, to which his curt reply was "Well we haven't", his answer being drowned by the noise of the angry fans outside on Brunshaw Road chanting "Sack the Board". Just two years after the great escape there was serious unrest yet again and the team was on the slippery slope, still in the Fourth Division.

A bad day was made even worse that afternoon because of the Hillsborough Disaster unfolding sixty miles away in Sheffield. It was the darkest of days for football in this country when ninety-six people were crushed to death at the FA Cup semi-final between Liverpool and Nottingham Forest. The Leppings Lane end of the ground, where I had stood exactly fifteen years earlier watching Burnley play Newcastle in their semi-final, had now become a morgue. This was a day destined to change the design of football stadiums forever as all-seating was to replace standing terraces and crush barriers.

We drove away from Turf Moor that afternoon with a sickening feeling in our stomach, but was fate also playing its part on my own club? Would I ever see a revival or was the massive scare of two years earlier actually just a rehearsal for worse to come?

"Looking back, there's no point denying the disappointments outweigh the moments of magic," said Alistair Campbell a few years ago. I had to agree.

Living forty miles north of London on regular shift work there was little enthusiasm or opportunity to watch the team I had supported since 1960. But the old cliché about Burnley being the first result I listened for still held

My November 2007 meeting with Andy Lochhead, Burnley's outstanding centre forward in the 1960s.

true, and then after seven dark years in the bottom flight there was the first ray of sunshine. Manager Jimmy Mullen crafted a side to win the Fourth Division in the first week of May 1992, just the start of the revival which saw another promotion via the play-off final against Stockport at Wembley just two years later.

Football was not quite at the top of my agenda by this time as our first boy Charlie was now eighteen months old with George on the way. If anyone tells you that children do not change your life then don't believe them. For Lynn and myself our family was more important than anything, even Burnley FC.

Sadly, poor performances and plummeting crowds brought relegation again after only one year in the new First Division and history was in danger of repeating itself. Despite several managerial changes it was not until Stan Ternent, a former Burnley player in the sixties, was appointed that this so called 'sleeping giant' could aim to climb back up the League.

Stan was a man who 'wore his heart on his sleeve'. He lived for the Club and kept his home in Burnley despite positions at other football clubs around the country. He even wrote to me on a couple of occasions to check my availability for charity golf tournaments – little did he know I couldn't tell my three iron from my sand wedge! Even so I made a short appearance on 'Football Focus' one Saturday afternoon pledging my support for Stan and suggesting that if anyone could get the team going it was him.

I was also appointed president of the Southwest Clarets so I could get information from the travelling supporters who followed the team week in, week out.

**Ian Wright shows off his new colours in February 2000. The former England and Arsenal legend was without doubt a major factor behind Burnley's promotion back into the First Division just three months later. (EMPICS Sport)**

It was a sticky start for Stan Ternent but promotion back into the First Division was achieved on 6th May 2000. Without doubt the introduction of Ian Wright, former England and Arsenal legend, was a major reason why success occurred as it did. Friends from their days at Crystal Palace, this was a fantastic coup which the Press lapped up. For me and thousands of supporters I felt this was a remarkably astute short-term signing and another promotion meant that Burnley could start the new millennium just one rung below the Premier League.

Sadly season 2002-03 proved a major embarrassment despite finishing clear of relegation. Defeats 5-6, 4-7 and 2-7 were more like rugby scores and over the season we conceded a massive 89 goals, even more than the relegated teams.

Stan was eventually relieved of his duties at the end of the 2003-04 season after six 'interesting' years. Now under the guidance of modern-day tracksuit managers Steve Cotterill and more recently Owen Coyle there have been great strides forward.

It would be over eighteen years before I was again invited by the Board to go behind the scenes of Burnley Football Club. It was Owen's first match in charge against Stoke City on the 24th November 2007 on a typically dark afternoon in East Lancashire with rain slanting across the ground. Burnley ditched their traditional claret and blue strip for the original pale blue and white stripes to commemorate their 125 year history.

Sadly the game ended 0-0 but the day meant more than the result for me. Having first watched my idols over the perimeter wall on that Good Friday in 1960 I spent much of the afternoon in the company of Andy Lochhead, a former centre forward and goal machine. At half-time I was introduced into the centre circle by Darren Bentley, the commercial manager, where I acknowledged the loyal supporters and met a more recent Burnley hero in Glen Little.

Almost fifty years of drama and heartache, brilliant goals and superb players have graced the pitch since I first got my dad to book our tickets on the football special from Todmorden.

I felt very humble, but so proud, as I looked out at the new generation of supporters who like me anticipated football at the very highest level in the near future, reminiscent of those glory years when I too was just a schoolboy.

My final thoughts have to be about last season, 2008-09, because the team and the management can hold their heads high after what was achieved. Beyond the euphoria of the cup games came the climax of the Championship season itself which had begun with two defeats but ended with a passage into the play-offs.

Burnley played Sheffield United on 25th May in front of 80,000 spectators at the New Wembley and this time I made the journey south to enjoy the spectacle. After a warm but cloudy morning the sun broke through at midday as I walked the short journey from Wembley Park station along Wembley Way towards the stadium. The atmosphere was already fantastic with 36,000 Burnley fans having made the journey from all over Britain and one even flying in from Mexico City. There are a few disadvantages about being in the public eye but on this momentous day I was happy for people to stop and chat, even though a few thought I worked on the Antiques Roadshow.

It has often been said that you never see Eric Knowles and John Kettley in the same room together!

I had not seen Madeleine Baker since we were at school together in 1970 but there she was with her husband as we re-lived the past for just a few minutes, wondering if our good mood would continue until five o'clock.

I have to say that both sets of supporters mingled without a hint of trouble as far as I could see, all a far cry from those dark days of football violence so common throughout the seventies.

Once inside the ground I met other people I had not seen for ten years and many more who knew me from television and we all chatted as though we were long-lost pals. One group of girls had come across from Valencia in Spain sporting their Burnley shirts and it was quickly becoming a day to cherish as the drinks flowed.

The hot sun shone down onto the pitch as the game kicked-off at three

Jubilant celebrations at Wembley when Burnley beat Sheffield United in the League Championship play-off on 25 May 2009. Some 36,000 fans witnessed the spectacle. (AP/Press Association)

o'clock under the open roof. After just thirteen minutes the Burnley end of the ground erupted as Wade Elliot scored the goal which would see the club return to the top flight in English football for the first time since 1976.

By the final whistle ominous clouds were gathering, but only at the Sheffield United end of the ground. For the 'clarets' the match was won and thousands enjoyed the moment with a tear of emotion in their eye, me included.

When it was first announced in November 2007 that Owen Coyle was to be the new manager I received a call from the *Lancashire Evening Telegraph* for my views.I had to be honest and say that I knew little about him apart from his recent good work at St Johnstone but I was delighted a successor had been found who was young and hungry.

"Maybe he will be the next Sir Alex Ferguson," I quipped.

Two years on and perhaps that was a good call. Maybe he will reach the heights of football management – there is no doubt he has made a magnificent start to his career south of the border – but knowing what I know now he would prefer to be 'the first Owen Coyle' with his own style and fresh ideas.

## 5.

# CHARITY CRICKET – LORD'S TAVERNERS AND MORE

Surrey and Yorkshire dominated our domestic cricket in my youth and like thousands of enthusiastic cricketers I had sporting dreams.

Dreams of playing for Yorkshire, dreams of scoring a century at Lord's and dreams of representing England.Of course I was never good enough but anything is possible in your dreams and as my heroes had done all those things surely those aspirations were impregnated on my mind! Among them were Fred Trueman, the greatest of them all, with a bespectacled young opening batsman named Geoffrey Boycott just bursting onto the scene. Sadly the realisation was that the closest I would ever get to meeting my cricketing legends would be in the cheap seats at Headingley, but a chance move into radio and television helped bring us together.

Imagine my surprise when Mike Gatting, Middlesex and England, wrote to me out of the blue at the BBC inviting me to write a few words on his behalf for his Benefit magazine in 1988.Why has he thought of me? What can I offer from a weather perspective? Did he have the right man?

Apparently his young children enjoyed watching me on the television presenting weather forecasts and it was no secret that I mentioned cricket at every opportunity, not least on a Saturday afternoon just before Grandstand.The seeds were sown – my interest in sport and cricket in particular had been noted by none other than the current England number four batsman and captain.

I duly wrote a piece about my Yorkshire roots and how I had scored centuries as a young boy using a rubber ball and wickets chalked on the factory wall, the cobbled street also making my 'quicker one' virtually unplayable as it deviated off the cracks on the ground.Clearly I had created such a good impression that Mike then sent me a fixture list for his forthcoming benefit season inviting me to play a match if convenient in my busy schedule.

Sunday 21st August 1988 was the date which was to impact on the rest of my cricketing life and also bring exciting career opportunities my way in the next twenty years. Opportunities I could never have imagined, even in my wildest dreams. I presented the weather forecast on TV that Sunday lunchtime then ran out to the BBC car park and drove off to Enfield CC in North London where the match was due to start at 2pm. A quick bite to eat in the marquee and my first chat to Mike Gatting on his home ground. An attractive setting with trees around the ground and obviously the weather was kind, fair weather clouds punctuating copious amounts of blue sky.

"Well John, do you bat or bowl?" Clearly he had never read my article!

Tentatively I replied with "a bit of both but not very well these days, my best years are already behind me. In the past I would say my bowling was more reliable".

"Right you can bat at number six. We've won the toss."

I have to say this made me even more nervous and my feelings turned to terror as his Middlesex X1 collapsed to 25 for 4 – I was in next! Despite my best efforts it was soon to be 45 for 5 and my opportunity to impress with the 'willow' disappeared as quickly as it came, scoring seven runs along the way but experiencing a partnership with one M Gatting, England captain, at the other end.

"Good effort," said Mike as I trudged to the pavilion, "better luck with the ball."

At least I could then relax until after tea so it was a great opportunity to chat with some of my team-mates including the late Wilf Slack, Keith Brown, Roland Butcher, Steve Gatting, Andy Crane and Henry Kelly.

Now Henry was not a man to use five words when fifty would do! He talked and I listened but occasionally we did reflect on our respective cricket interest.A jovial Irishman who was more familiar in the surroundings of a TV studio as one of the presenters of 'Game for a Laugh' – perhaps a suitable title for our cricketing abilities so far that summer afternoon!

After tea it was the turn of Enfield to bat whilst a plethora of media people and proper cricketers took to the field in the hope of Gatting's team achieving yet another victory in his benefit season. For me the hope was that I would get a bowl sooner or later but not until the amateurs from Enfield had negotiated several overs of pace from the likes of Norman Cowan, another England cricketer on show that day.

After about twenty overs I was called upon at the pavilion end to show this corner of North London what I could do, hoping not to trip over my boot laces along the way. I marked my run, some twelve paces in those days, and delivered my first ball. All was fine, in fact the over went well, spurred on by enthusiastic fieldsman gathered around the wicket keeper. My relief at bowling a couple of decent overs then gave me the confidence to try and bowl a bit quicker, maintaining line and length as every doyen of the game would insist.

No doubt the adrenalin was pumping as my desire to take a wicket increased and it was a huge relief when my 'straight one' flattened the batsman's leg stump in my third over. I could have been bowling for England at Lord's (was I still dreaming?). No matter, the thrill of taking a wicket when called upon by the England captain at his home club was fantastic. Another wicket was to follow but after six or seven overs I got the call that no bowler ever wants to hear, "Have a break, John."

In the end the game was won and I had not let myself down but more than anything else it was my passport into the famous Lord's Taverners. Their patron and twelfth man is the Duke of Edinburgh and the Taverners was founded at an inaugural meeting on 3rd July 1950 as a club for members of the arts. The entrance fee of £1 was agreed, to be donated to the National Playing Fields Association, Prince Philip being their president. John Mills (later Sir John Mills) was invited to be the first president of the Lord's Taverners. 1952 saw the first three charity cricket matches being played at Bishops Stortford, Hayes and Stratford upon Avon.

So impressed had Henry Kelly been that he asked if I would like to join the Taverners. On my behalf he duly organised an application form inviting me to join the ranks of the great and the good. The form was surprisingly succinct and straightforward but it did ask the applicant what they could bring to the Taverners, in other words did you have any special virtues or skills. My response was that I could guarantee excellent weather at every outside event such as cricket or golf matches. In addition I could speak for one minute or two minutes and finish exactly to the second without deviation or repetition! My entry form was seconded by Henry Kelly and I became Lord's Taverners member number 2408.

Having surprised myself by taking two wickets at Enfield for Mike Gatting, I was anxious to see whether I could now scale the heights and manage to get a game for the Taverners in my first year of membership, the 1989 season. As we were living in North Hertfordshire I made myself available for a local match at Fenners, home of cricket for Cambridge University. I had actually seen Mike Atherton score a century on the ground when he was captaining the side during his university days but I knew the pitch had a reputation for being 'flat'– more suitable for scoring runs that taking wickets – but that mattered little as long as I was selected.

My request for a position in the team was accepted and I made my debut at Fenners against a side from East Anglia consisting of local radio and television stars and good standard amateur cricketers. Lunch in the pavilion before the match was incidental to the reason I was there, to meet and play cricket with people from the world of professional sport and show business in order to raise thousands of pounds for charity.

The late Willie Rushton was a stalwart of the Taverners and very much 'old school', taking to the field in cream shirt and flannels with a tie wrapped around his waist. His athleticism had deserted him but his passion for cricket and enjoyment of the occasion was infectious. Sir Jeffrey Archer – MP and author, Julian Wilson – racing journalist, Nicholas Parsons – broadcaster, and Frazer Hines – actor, were other notable amateurs on show but the professional cricketers read like an Old England eleven. Former

**Debut match with the Lord's Taverners at Fenners, home of cricket for Cambridge University.**

captain Mike Denness, fast bowlers John Price, John Lever and Fred Rumsey, batsmen Roy Virgin and David Steele.

As was customary, the Lord's Taverners invited the opposition to bat first and the former England quick bowlers turned back the clock for almost an hour before the rookie bowlers were given a chance to shine. Unfortunately I failed to take any wickets in my debut match but on a glorious occasion in warm sunshine it mattered little. The crowd were sympathetic to the loose deliveries, dispatched by the batsmen to all corners of Fenners, and I was aware that it was a friendly wicket for the batters and if I did get a chance to bat it could be fun. Of course I had to wait a while for my turn but eventually I strode to the wicket, bat in hand. Not exactly brimming with confidence perhaps but hoping the bowling would be friendly with no hidden demons in the pitch.

A shaky start as usual (I have always been vulnerable in the first ten minutes of an innings ) but the confidence-boosting four through the covers gave me the springboard I needed. Suddenly my nerves disappeared and the benign pitch provided several more loose balls to tempt this enthusiastic amateur. I think 42 was my final score thanks to the charity of the opposition from East Anglia and we won the game, all the players being rewarded with a bottle of locally made white wine to remember the occasion.

Once again I was in dreamland!

It was with great pride that I continued to play several games every season for the Lord's Taverners with many highlights along the way. I always said I would continue to play until I didn't do myself justice but it was always understood that the only reason I actually got a game was to keep the rain away! Sadly that is easier said than done and I vividly remember a match in the grounds of St George's School at Windsor Castle which was brought to a sudden end after twenty minutes as a massive thunderstorm ground to a halt over the Castle and that was it for the day. It is naturally frustrating for the crowds and sponsors who aim to make it a very special day but I have to say just as bad for the players who give up their time in a good cause.

Arundel has been a regular venue for visiting touring sides for as long as I can remember but I too have a vivid memory of the one match I played there in 1996.

August Bank Holiday's reputation again contrived to ruin our match as heavy storms broke out after morning sunshine. Despite my very best efforts lunch continued much longer than normal with many of my fellow Taverners, including myself, doing a turn at the microphone to keep the diners relatively happy. At long last a shortened game began in glorious warm sunshine about four o'clock which eventually led to my only Man of the Match award.Three wickets on a helpful pitch followed by a half-century with the bat against some friendly bowling from John Barclay – Sussex, and Ross Edwards – Australia, made one weatherman extremely happy. A

Lord's Taverners' fixture in the grounds of St George's School at Windsor Castle, 2000. Seated in the centre is the Duke of Edinburgh, Patron and Twelfth Man.

presentation of three bottles of quality red wine from Sir Colin Cowdrey was the icing on the cake and I just hoped the crowd was as happy as I was that evening, especially as my family was also with me for the weekend.

Michael Holding was a West Indies fast-bowling legend happy to represent the Taverners when we played at Bury St Edmunds. He was always keen to play in any match within twenty miles of his beloved Newmarket where he has enjoyed many successful race days.

Imagine my delight when the commentator made the following announcement: "Replacing Michael Holding at the town end is weatherman John Kettley." For me that kind of information on the loudspeaker at a charity cricket match makes it all so worthwhile and made me feel very proud that I was still fit enough to perform.

My final game at Marlow, Buckinghamshire in July 2005 was a fitting climax when a moderate performance with bat and ball culminated in the 'Champagne Moment'. Fielding at mid-on I ran back towards the boundary and held a catch with one hand – pure fluke of course but once again a bottle was coming my way.

**John Major signs his autograph during a 1992 charity match at Alconbury for Norma Major. Lynn is pregnant with Charlie.**

I was also privileged to play in other charity cricket matches, notably for Mrs Norma Major at Alconbury. It was a balmy spring evening in 1992 when I was first approached at Huntingdon races to see if I would represent Mrs Major on behalf of Mencap.

So on a perfect afternoon on 24th July, I made my debut under the guidance of former England and Middlesex spin bowler Phil Edmonds. Equally famous colleagues for the day included Jimmy Greaves, Steve Smith-Eccles, Rory Bremner and Joel Garner.

David English captained the opposition as usual and I was to be given a fairly ripe 'cherry' to pit my wits against this dashing left-hand batsman. What came next was near fantasy as English lofted my straight but tempting delivery to deep mid-wicket only for a slightly overweight Gary Newbon to reel back the years and with amazing athleticism launch himself at the projectile before it reached the boundary. This was surely one of the greatest catches ever taken in these annual cricket matches and people (well Gary) have talked about it for years. Obviously the wicket-taker was overjoyed and it encouraged me to take a further two wickets, though I had to beg heavyweight boxer Gary Mason not to hit me!

I returned to Alconbury in 1995, 1996, 1999 and 2001, every time in perfect weather. Not surprising really because this is just about the driest area of the British Isles with an annual rainfall of about 550mm, less than 22 inches. Mind you in 1997 about two inches of rain fell in 48 hours to reduce the Recreation field to a bog and so this annual event went the same way as many others, including a washout at Wimbledon.

My cricketing days may be just about over but I have enjoyed the company of many wonderful people over the years, including Mike Gatting who was first to launch me into the world of celebrity cricket.

**6.**

# MORE EXTREME
# WEATHER EVENTS

In our temperate latitudes what precedes a major weather event, or follows later, may have no connection to the meteorological sandwich in the middle. The diversity of our weather helps to keep it interesting and never far from headline news.

The situation in 1975 and 1976 was a case in point because for many people these were notable dry, hot summers and will always be remembered as such. Interestingly the first of those hot summers began in bizarre fashion as followers of cricket will remember. Easter that year had been colder than normal with more snow across the country than the previous Christmas, a statistic which continues to stand the test of time. On 2nd June 1975 a cold blast of Arctic air swept through most of Britain and there were even reports of snow falling during county cricket matches, most notably at Buxton where Lancashire and Derbyshire were in opposition.

Dickie Bird was one of the umpires that day and he is a man known for attracting strange twists of nature when he is on duty. Many years later he had problems at Old Trafford during a Test Match when bright sunlight was reflecting off an open window and on another occasion at Headingley in 1988 he had to deal with a gushing drain on the outfield. It was the Test between England and West Indies which had amazingly started on time following heavy rain on the eve of the match. After just four balls the West Indies quick bowler Curtly Ambrose showed Dickie where water was oozing over his boots half-way along his run-up. The umpires ordered the players off the ground in bright sunshine leaving ground staff to discover that a drain had burst. Running repairs were carried out for almost three hours, and copious amounts of sawdust were needed to soak up the water.

Back to 2nd June 1975, the freak fall of snow which steadily accumulated on Buxton cricket ground led to the inevitable headline 'Snow stops play'. In the passage of time people recall maybe six inches of snow lying on the ground by the end of the day although in reality the depth was probably half that. For Dickie Bird there was no choice but to abandon play for the day and this remarkable 'summer' snowfall would always be remembered by the Lancashire and West Indies captain on duty that day, the great Clive Lloyd. He may never have seen snow before and for it to cause a cricket match to be abandoned during the summer must have been like living a dream for someone growing up in far-away Guyana.

For the record the maximum temperature in Buxton that day was 6.5C but just three days later the cricket World Cup began in a heatwave and much of that summer was to remain very warm and dry with an escalating drought in the subsequent twelve months.

The summer of 1975 was the warmest since 1947. After the extraordinary start to June, warmer conditions quickly extended northwards with afternoon temperatures getting into the high 20s Celsius for much of the remainder of the month, although there were short spells of cooler weather in mid-month and again over the last few days. Southampton reached 30C on the 12th but despite the drop in temperature at other times amounts of rain across the south proved small with parts of Wiltshire and Poole in Dorset having precious little.

July saw two periods when the Azores high-pressure system pushed northwards to lie across the British Isles for days at a time bringing hot and sunny conditions, notably in the first and last weeks of the month. Such is the nature of a British summer that the familiar saying 'three fine days followed by a thunderstorm' proved very appropriate as low pressure from Biscay triggered storms and some of the wettest weather of the summer during the second and third weeks.

Early August continued as July ended with high pressure in command and just a few thunderstorms to break the pattern of the summer warmth. Temperatures continued to soar and on 5th August the temperature peaked at 31.2C at Marham in Norfolk only to be surpassed with the 34.2C at Heathrow Airport three days later. During this spell new records were set with Glasgow having its highest temperature ever, 31C, reliable temperature measurements having been recorded there since 1868. By the middle of the month more unsettled, cooler conditions had returned from the west across the whole of the country, although there was a return to fine, sunny weather for a few days near the end of the month. It turned out to be the warmest August on record, eventually being surpassed twenty years later in 1995.

The Hampstead Storm broke in the late afternoon of 14th August and as I was working and living in Farnborough, Hampshire the event is still clear in my mind. It had been a very hot day and looking northeast from the house I was sharing at the time, the distant blue horizon was punctuated by tall, billowing white cumulonimbus clouds symptomatic of thunderstorms. Fully thirty miles from London it was clear that something nasty was brewing and the crackling on the radio confirmed that electric storms were already developing.

For almost three hours Farnborough continued to enjoy the hot evening sunshine but over Hampstead the storm was raging.

Eye-witness reports at the time said the rain began around 17.15 and continued until 19.50 with the storm centred over Hampstead Heath and barely moving the whole of that time. New cumulonimbus cloud cells were

**Flooding on the railway line at Todmorden in 1975 – a year that also saw what was then the warmest August on record.**

generating over northwest London as old ones decayed to produce a self-perpetuating storm. Rain gauges in the area suggested that as much as 170mm (almost 7 inches) of rain fell in less than three hours and the repercussions of that massive storm would affect millions of workers and residents across the capital.

Four main-line railway stations in London were flooded and closed with a substantial part of the underground network also put out of operation as tunnels flooded. It was one of the largest amounts of rain to have ever been recorded in Britain in such a short time and was accompanied by the proverbial 'marble-sized' hailstones. All in all the Hampstead Storm was typical of how a spell of hot weather can come to a sudden and dramatic end.

After some useful rain in September much of the autumn and following winter continued drier than normal so much of the country was already facing a situation where rain would urgently be needed in the spring. Otherwise the dreaded hosepipe ban would quickly be top of the agenda with the summer just around the corner. It requires steady winter rains to reduce the soil moisture deficit and replenish the reservoirs because summer downpours from localised thunderstorms only superficially wet the ground and most runs away without adding usefully to fields and gardens.

However one of the worst gales to hit Britain happened during that dry winter on the evening of 2nd January 1976. For thousands living close to the east coast the development was similar to the infamous North Sea storm surge of January 1953 when hundreds perished in widespread flooding. Once again a deepening depression crossed Scotland and moved southeast down the North Sea but thankfully on this occasion it coincided with a low tide and not the very high tides experienced twenty-three years earlier.

Severe gales were widespread across Britain that night with exposed places like Cromer on the north Norfolk coast battered with winds gusting as high as 108mph – just a little short of the highest achieved on the south coast in the 1987 storm. Norfolk was probably worst hit and at one time Norwich was cut off as every road became blocked by fallen trees and it was said that perhaps 600 trees had come down in the city itself. Damage to property was extensive in many areas of the country but the track of the storm left a swathe from Northern Ireland to East Anglia worst affected. As in any major storm, power lines were down, rail travel was seriously disrupted especially through the Midlands, and every town and city was littered with fallen masonry. As it happened the improved sea defences along the east coast following the devastation of 1953 played their part although Cleethorpes saw one section of the wall breached. In all it was estimated that nearly thirty people died across the British Isles as a result of road accidents and falling trees attributable to the storm.

Apart from the storm it turned out to be a dry January followed by a drier than normal February and March. By April 1976 the situation was clearly worsening as the rain almost dried up in many areas. A sprinkling of thunderstorms sparked off during the increasingly hot May, when as early as the 7th the temperature at Waddon, near Croydon, had already peaked at 29C – still a record high temperature for that date!

Although there was still a lot of dry weather in June my memory of it was that it did turn cooler and cloudier during the first three weeks of the month, for which I was eternally grateful. Now in Coventry sitting my final year exams it would have been too great a distraction had the weather been too hot and stifling. So luck was on my side as I tried desperately to achieve my goal of an applied physics degree so I could get back to the Met Office and pursue my aim of becoming a weather forecaster.

Just like switching on a tap the weather suddenly turned hot again as I left Coventry in the third week of June. For me it was six weeks between university and returning to work for twelve months at the Meteorological Research Flight back in Farnborough.

Now it was the start of the Wimbledon fortnight and the most arid championships ever. As long as I can remember the start of Wimbledon has generally been seen as a rainy spell with the perennial image of the ground staff bringing on the covers at the slightest threat of a shower. "Ladies and gentlemen, play is suspended" is the familiar call from the umpire's chair. But not in 1976.

Not only did it stay dry for the entire fortnight but it was very hot and oppressive throughout. It seemed that just a couple of days had passed before the parched courts were turning brown as the ground continued to dry out under the 'tropical' sun. My old friend and colleague Philip Eden once suggested that the Wimbledon fortnight in that memorable year could well have been the hottest fortnight ever in the London area whilst records have been taken. He suggests at least since 1783 and probably for much longer!

Observations from nearby Kew Observatory indicated that for every day of play the average afternoon temperature was an incredible 31C (88F) and actually peaked at 34C on both Saturdays. The sun shone on average for twelve and a half hours every day and for thousands of spectators not able to get under cover the fierce heat must have been unbearable. Indeed the medical staff on duty had to deal with scores of people with sunstroke or dehydration. At least the players could have a break at the change of ends and take liquid refreshments with the unusual benefits provided by the spare ball-boys and ball-girls of an umbrella held over them as protection from the hot sun.

This was a remarkable period of summer weather in this country and particularly at SW19, often maligned for the rain but this time truly memorable.

Elsewhere the temperature reached 35.9C at Cheltenham on 3rd July and it came as no surprise when the water authorities announced water restrictions and we were all encouraged to recycle bath water and use it in the garden. During the heatwave hospital admissions soared as millions suffered from the heat with sunstroke and heart attacks increasingly common. Most bizarre of all was the appointment by the government of a Minister for Drought! Former football referee Denis Howell was the man called upon to sort out the weather, or at least give advice and make recommendations how the country could survive this crisis – said to be the worst since World War Two. An unenviable task made worse as July turned to August.

By now I was back in the south of England in an area where heathland and woodland fires were escalating and stretching the emergency services to breaking point with the army now being called upon to help fight the fires. Not a day passed when there wasn't another fire breaking out, not only across Surrey and Hampshire where I was but in many other parts of tinder-dry Britain.

The drought was becoming more severe through every passing day and the warm and oppressive nights for those living in the cities, the urban heat islands, meant that sleeping became almost impossible. Out in the country it was less of a problem as the parched ground allowed the temperatures to dip dramatically overnight. Across much of the south the August sunshine

**Water standpipes provide a fitting foreground for Denis Howell, bizarrely appointed Minister for Drought in the scorching summer of 1976. (PA Archive)**

totalled up to 330 hours with daytime temperatures regularly between 25-30C but with the anticyclone now settled from Ireland to Scandinavia there was more of an easterly wind which led to the hottest weather being further west.

Throughout the three summer months of June, July and August there were 62 days when the temperature somewhere exceeded 25C – and all of this before global warming was talked about. Indeed this was at a time when some scientists were even contemplating the next ice age as they had noted a drop in global temperatures!

For the vast majority of the British Isles there was nowhere to hide from the intense heat in 1976, although the position of the jet stream continued to bring rain and cool conditions to the west of Ireland, northwest Scotland and the Hebrides. For that area it was certainly not a memorable summer. However for the rest of us a ban on the washing of cars and the use of hosepipes had been in force for months and some rivers were either dangerously low, or in sandy or chalky areas, dried up all together.

On 25th August, shortly before the Late August Bank Holiday, a rain-making guru from India along with his circle of priests were seen on television in London. Coincidentally, though a minority may have other ideas, the next few days saw the rain return. A thundery disturbance in the upper atmosphere over eastern France chose to move north and the first thunderstorms appeared in East Anglia over the weekend. Ipswich was one of the first places to catch the storms and flooding was inevitable as torrential rain fell onto hard ground with rapid run-off. King's Lynn also suffered badly and by the early hours of Holiday Monday there were reports of over four inches (110mm) of rain in the area. The weather quickly changed over the rest of the country as the welcome rain arrived - then it rained for week after week leading to one of the wettest autumns on record!

Soon the exceptional drought and the two remarkable summers were distant memories and water restrictions gradually lifted but the fact that they are still so clear in the minds of millions in Britain shows just how rare the whole event had been.

The events during the winter of 2008-09 served as a stark reminder that our climate still lends itself to harsh weather of a more traditional type. During the 1950s and 1960s Britain suffered a period of cool summers and 'normal' winters, apart from the exceptionally severe winter of 1962-63, when heavy snowfalls would come along periodically through the coldest months of the year, lasting for a week or so before milder westerly winds brought a return to rain and gales. The winters of the early 1970s were uneventful, except steady snow in the early hours of Christmas Day 1970 brought an official white Christmas even in London.

Only two years after the hot summer of 1976 the country saw extreme snowfalls and blizzards once again. During the winter of 1977-1978 conditions became very severe on occasions.

The Highland Blizzard of 28th January 1978 resulted from a deep depression which slowly tracked across central England leaving strong north-easterly winds over Scotland. Power supplies were soon lost in the blizzards across northern Scotland and it was a heavy fall of snow with severe drifting that brought transport links to a standstill. One man survived being buried in his car for eighty hours but three people died in the bad weather.

February 1978 was a very snowy month for northern Britain along with the southwest and Wales. By the 11th snow was lying to a depth of 30cm in Durham and the Isles of Scilly had their first snow since 1963.

Heavy snowfalls continued to sweep across Devon but a deepening depression lying off the southwest coast on the 18th allowed cold easterly winds from the continent to intensify the storm as the contrast in temperature between the two systems increased. This brought very heavy snowfall to the whole of the West Country and east as far as the South Downs.

This latest episode became known as the 'The Great West Country

Blizzard' and brought transport chaos to the area as gale force winds piled up the snow. Reports at the time suggested that exceptionally heavy snow fell for up to thirty hours leaving drifts up to nine metres deep across Dorset. Weymouth and Dorchester were completely cut off by road for several days with rural parts of Dorset isolated for up to a week. Cardiff also reported snow to be lying to a depth of 30cm or more. One report from Dartmoor at the time talked of a "freezing white hell and a screaming wind that tore at us in all directions. We were literally stopped in our tracks, unable to think or walk or speak or see".

The following winter, 1978-79, was even worse and actually the coldest since 1963. It was also the so-called Winter of Discontent when industrial action was an epidemic which left the country paralysed as power cuts became commonplace, schools closed and rubbish left in the streets was uncollected for weeks. Despite the bad weather the appearance of a gritting lorry became a rarity as strikes escalated with the result that road travel during the worst of the weather became treacherous and even impossible on occasions.

As with most harsh winters it was not until late December that conditions suddenly worsened. Waterlogged fields quickly froze as temperatures tumbled, and heavy snow was whipped up into deep drifts by the strengthening Arctic winds that swept across the south of England. New Year's Day set the scene for what was in store throughout the month with temperatures remaining below freezing all day across most of England. January 1979 remains one of the coldest on record across the British Isles with a succession of cold and not so cold days, frequent frosty nights and bands of snow. Throughout the winter period even Brighton on the south coast had seven days when the temperature failed to get above freezing but that paled into relative insignificance compared to the twenty-seven days it endured back in 1962-63.

The middle of February was also severe, East Anglia suffering more than most as the Siberian winds brought blizzards and heavy snowfalls from the northeast. On the 14th and 15th road travel ground to a halt as conditions rapidly deteriorated and power cables crashed to the ground under the weight of the snow and the severe gales. Hundreds of trees blocked the roads and by the 16th Ipswich was described as a 'ghost town' as a thick blanket of snow and severe drifts covered the whole of East Anglia. Motorists were left stranded 'all over the place' according to the RAC and milk and beer deliveries were cancelled until further notice. By the following day there were still no buses running in Ipswich and for five consecutive days temperatures remained below freezing. It would be a very slow thaw for the rest of the month into early March but roads remained treacherous after re-freezing again overnight.

Buxton in Derbyshire also suffered from several snowstorms during that

**The winter of 1978-79 was the coldest since 1963. Home-going commuters had little to cheer them on the evening of 15 February 1979 as a blizzard swept south and paralysed public transport. (YPN)**

harsh winter. Blizzards in the third week of January virtually cut off the town and winter's icy grip brought repeated falls of snow for several weeks. For some living in the Peak District around Buxton this was as bad as the winter of 1962-63 with abandoned cars littering the National Park and sport badly hit. The local football club could not play a home game for fully ten weeks due to a combination of ice and snow. A light at the end of the tunnel in early March was soon extinguished by renewed snowfalls on the 16th and 18th and another freeze which continued for over a week.

A few more hard winters followed into the early 1980s by which time I had moved north again to work at Nottingham Weather Centre, based in the village of Watnall on higher ground six miles to the northwest of the city. On at least one occasion I remember walking over five miles from my new home at Brinsley through deep snow to start my shift. In those days it was expected that you would get to work, even when the roads were blocked and public transport was never an option. Clearly exhilarated by a fresh fall of snow it was also imperative that the previous nightshift was relieved after being on duty for twelve hours. As Ian McCaskill found on the morning of the Great Storm in 1987 there is no feeling to match that of expecting your colleagues to arrive for work …. but no-one walks in!

Whatever the weather, and there was plenty of it about in the early 1980s in the north Midlands, you got to work.

It was around this time that the coldest recorded temperatures occurred in Britain and it became the norm to wake up to ice frozen on the inside of the house windows. December 1981 was the coldest December for over one hundred years for many! Without question it was also the snowiest December of the century and for most of the month the ground was deep in snow.

It was not until the start of the second week that temperatures had dropped below freezing in the south but a heavy fall of snow around the Chilterns was a sign of things to come until a temporary thaw at the month's end. Regular bands of heavy snow kept adding to the problems of travel across the whole of the British Isles but between the blizzards, when the winds fell light and skies were clear, the temperature dipped alarmingly over the snowfields.

On 13th December a new record minimum temperature for England was established at RAF Shawbury, Shropshire, when the thermometer plummeted to minus 25C (-13F) following a recent fall of snow. The previous day the temperature never exceeded minus 12C (10F) because of persistent, dense freezing fog. This record was to last only twenty-eight days because the second week of January was permanently frosty and this time, on the 10th, nearby Newport went even lower to minus 26.1C (-15F). Not wishing to be outdone Braemar in Scotland equalled its own record low UK temperature of minus 27.2C (-17F).

Bone-chilling hardly does justice to the severe cold of that winter but few people will remember it as they do 1947 or 1963.

More memorable for people in the southeast were the bitterly cold conditions and heavy snow which hit London in January 1987. It was a classic scenario of dry, cold Polar Continental air sweeping snow showers into eastern areas as they picked up moisture from the North Sea. Initially a deepening low pressure system moved south across Norway into Germany which was followed by a surge in pressure over Scandinavia. Intensely cold air was now spreading from western Russia into Poland and Germany and it had our name on it!

By Saturday afternoon, 10th January, eastern England had seen temperatures fall below freezing and this icy weather extended across the whole of the country within twenty-four hours. I presented the Countryfile weather forecast the following day on BBC1 which included the memorable phrase: "The only bright thing on this forecast is my tie!" Computer models were now predicting freezing temperatures and blizzards all week and that

**Frozen canal near Derby during the period when 1981 brought the coldest December for more than a century over much of Britain.**

was the message I had to get across to the public.

The first snow showers had first appeared on the Saturday but by Monday 12th conditions had deteriorated dramatically and it was to be a day off school for thousands of kids across the southeast. A major weather story was unfolding which culminated in London having one of its coldest twenty-four hours on record. Across the region motorists were urged to stay at home as the emergency services had to deal with thousands of breakdowns and accidents due to the severe weather. Kent and Essex were worst affected with reports of as much as 60cm of snow falling overnight and British Rail cancelled over three hundred trains.

That Monday was one of the most remarkable weather days of the 20th Century as temperatures stayed below minus 5C throughout England. When you watch weather forecasts on television these days and you hear the presenter talking about 'bitterly cold winds' when temperatures are a few degrees above freezing, just imagine what a proper icy wind would have felt like in 1987. The wind-chill factor at that time was probably making the temperature feel more like minus 20C!

For several more days the country suffered from further snowfalls and freezing temperatures and it must have been during this period I was at the mercy of British Rail and the first indications that we could have been experiencing the 'wrong kind of snow'.

I finished my night-shift at the Weather Centre in High Holborn and trudged my way north to King's Cross station, watching cars slither along the eerily quiet streets of the capital as heavy snow quickly began to accumulate. Once on the train to take me to Letchworth, Hertfordshire, it would only be a few seconds before the train ground to a halt in the short tunnel just north of the station. It would be forty-five minutes before the train moved again and it was clear that this latest fall of snow was already having an impact on the transport infrastructure. What made it worse for me was the fact that stuck in the tunnel I couldn't even see the snow settling outside.

There were no further dramas on the journey, although the usual 50-minute trip turned into something like two hours or more with a further mile walk at the other end.

The Great Storm of Friday 16th October 1987 is rightly remembered by millions in southern Britain as one of the most severe weather events 'in living memory'. Nineteen people died early that morning but the death toll would almost certainly have been greater had the storm arrived in daytime.

Just over two years later the Burns Day Storm, so called because of its occurrence on the anniversary of Robbie Burns' birthday on 25th January 1990, did sweep across Britain in the middle of the day and more people were tragically killed as a result.

Being mid-winter the jet stream which drives weather systems across the Atlantic towards western Europe was in its normal position as opposed to lying well to the south in the Bay of Biscay. A deep depression, or low pressure system, crossed the south of Scotland close to Robbie Burns' birthplace in Ayrshire with a central pressure of 953 millibars. This was very similar to the depth of the 1987 storm but this time the track of the storm was critical and meant a far larger area of the UK was hit by the severe weather. It was also the middle of the day when the centre of the low had its greatest impact bringing widespread severe gales as millions of people were going about their daily business. The highest gusts fell just a little short of its predecessor, the highest recorded being 107mph at Aberporth, but other gusts in excess of 100mph were common to the south of the centre of the storm.

There were hundreds of casualties from flying debris and falling trees but sadly forty-seven people also lost their lives as a direct result of this daytime storm.It was well documented at the time that Gordon Kaye, an actor from television's 'Allo 'Allo, was lucky to escape with his life when an advertising board crashed onto his car. He was to suffer serious head injuries and also went into a coma but he did live to tell the tale. The death toll was the highest in any weather related event since the North Sea storm surge of January 1953. Perhaps it was a blessing that fifteen million trees had already come down in the Great Storm of October 1987 otherwise if they had chosen that January day to crash to the ground many scores of people

**Aftermath of the 'Burns Day Storm' of 25 January 1990, as seen in Baldock Woods, Hertforshire.**

going about their daily business could have become statistics. The insurance industry was hit by substantial claims as damage to property became a major issue yet again.

I had my own story to tell along with millions of others because at the height of the storm I was risking life and limb whilst hanging on to an ageing conservatory roof. Lynn and I were house-hunting at the time and living in rented accommodation in Baldock, Hertfordshire prior to getting married. The semi-detached house had seen better days but served us well – at least until that wild January day. In the end I took the decision to let nature takes its course and the glass roof was sent smashing against the trees at the back of the house. I dread to think what the final outcome could have been but this was only a rented house and I had no intention of becoming a statistic. A hero I was not but my broadcasting career and family life were intact.

In the final analysis it was a winter storm no more severe than those which sweep through northwest Scotland at least twice in most winters. From a media point of view those storms are nothing unusual and affect a much smaller percentage of the population. Damage to property in the Highlands and Islands is often negligible. But the track of the storm is critical and it is absolutely vital for forecasters accurately to predict their movement so that all the relevant authorities can be warned and the general public be on standby to do what they can to secure their own property. As a general rule the strongest winds are to be found to the south of the storm track.

It is also important to remember that the weather has no boundaries and no memory! Lightning can strike twice but equally a severe storm can wreak havoc in the same areas at least once – more of that later with recent flood events.

Four years later in February 1991 the rail network was crippled again by the unusual powdery snow which typically falls when the air is very cold. The problem seems to lie in the fact that when the snow is light it can easily blow into the engine workings of the electric trains and cause short circuits. In other words the units become inoperative and the network fails.

Media reports during the recent heavy snowfalls in February 2009 often referred to the fact that conditions had not been as bad in southern England since the events back in 1991. Those comments can be substantiated because the period from 5th to 12th February was indeed quite severe, even in London, as the streets were covered in snow throughout. However, I would say the overall depths of snow in the Home Counties in 2009 were slightly greater, although Hampstead measured 25cm of lying snow and St James Park 20cm on the 8th when conditions were at their most severe. Further afield snow depths were probably greater in 1991 with as much as 30cm of snow lying in Chelmsford and 50cm at Bingley in West Yorkshire.

A few years later the winter of 1995-96 was also noteworthy being the coldest winter over England and Wales prior to 2008-09, but interestingly hard on the heels of its summer drought. Some heavy snow struck the North York Moors during the third week of November which served as a taster for the winter ahead when the cold north-easterly winds made a reappearance after a few milder winters. After a largely uneventful first half of January 1996, when mild, cloudy south-westerly winds dominated, the winds suddenly switched around. The Siberian high pressure extended to the west over Scandinavia sweeping much colder winds towards Britain. By the fourth week there was a scenario reminiscent of January 1987 when even daytime temperatures struggled to get above freezing and the strength of the east wind brought penetrating frosts.

Snow showers turned heavier and more frequent in the east, quickly spreading west with a further heavy fall developing across northern England. The snow drifted in the strong winds but because the snow was only transient across southern counties the situation never hit the headlines. Nonetheless the cold anticyclone continued to dominate into February bringing severe overnight frosts. Battle lines became drawn to the west as rain bands advancing from the Atlantic were quickly turned to heavy snow as they met the dominant cold air. By the 7th February parts of Cumbria and southwest Scotland were cut-off by snow said to be lying to a depth of 50cm in some places and travel became virtually impossible. A state

**Severe conditions in the second week of February 1991 brought the rare sight of snow at sea level, with this family bravely venturing on the beach at Scarborough. (YPN)**

of emergency was declared in Dumfries and Galloway as scores of motorists were left stranded on the A74.

Very little of the snow got beyond central areas leaving the east virtually unscathed but at its height some of those western areas of the British Isles had probably witnessed their heaviest snowfalls since 1963, and possibly even 1947.

After a short respite another Arctic blast later in the third week brought widespread heavy snow showers and blizzards even into East Anglia, and the strong winds and high tides were also responsible for coastal flooding.

As we entered a new millennium the threats of global warming continued to escalate but the severity of our winters declined, with the obvious exception of 2008-09. Most would agree that the very severe winters are now confined to history but 'every wind has its own weather' and we should still anticipate that air with Arctic origins will prove the catalyst for further snowfalls in the future and the inevitable headline "roads gridlocked as two centimetres of snow settles at rush-hour".

**Trudging through the snow in the bleak midwinter of 1996, when penetrating frosts and the strength of the east wind added to the misery. (YPN)**

Clearly the prolonged dry spell of 1975-76 was quite exceptional but it is often forgotten that conditions more recently in 1995 also proved to be extreme. Ironically the winter months of December 1994 through to February 1995 had been very wet, perhaps the wettest on record, but by the spring high pressure was dominant and with the Azores anticyclone reluctant to move there followed another period where hosepipe bans were high on the agenda. For five months the rainfall was well short of long-term expectations and with hot weather during July and August the countryside was again showing the scars of drought. Unusually it was my home area of West Yorkshire which on this occasion suffered most as reservoirs fell to dangerously low levels and the only worthwhile rain from thunderstorms in the summer months was largely lost to run-off.

Statistically the summer of 1995 was actually slightly drier than 1976 in England and Wales, and a record breaker. The summer holiday weather was at its peak at about the same time as nineteen years earlier, from 19th June to 29th August, but the most intense heat occurred in the first week of August.

Although the rain returned through September there was insufficient to help improve water supplies across the Pennines and Lake District as the wettest weather remained in the south. October was then quite dry and mild followed by a cold but drier than average winter which all served to reduce some reservoirs in West Yorkshire to below 15% capacity. There was a continued dearth of rainfall for much of 1996 as well, so by March 1997 England and Wales had experienced its driest two year period in 221 years

**Unusually it was Yorkshire that especially suffered in the drought of 1995, with reservoirs falling to dangerously low levels. Cracked earth and the bridge from the 'drowned' village of West End told their own story at Thruscross in the Washburn valley. (YPN)**

on the rainfall series. Once again river flows were very low and in some cases, like 1976, springs even disappeared.

Britain has seen other such extreme examples of drought and heatwaves over the centuries, such is the natural variability of our weather, but these tend to be relatively short spells in the climate overview.

One such heatwave struck the British Isles during August 2003 when the temperature rose to unprecedented levels. Hottest of all was Faversham in Kent with 38.5C on the 10th, just one day after Scotland's highest ever temperature of 32.9C at Greycrook in the Borders.

For my family the hot weather could not have come at a better time because we were having a new kitchen fitted in our thatched cottage near Hitchin, Hertfordshire, which would take two weeks to complete. Not one drop of rain fell during that fortnight and my culinary skills on the BBQ improved with practice as we ate daily in the garden, sheltered from the stifling heat of the day under the gazebo. The joys of an English summer!

Sadly the searing heat brought severe health problems to many millions across Europe. It was estimated that the hottest August on record in the Northern Hemisphere had been responsible for the deaths of 2,000 people in the UK alone and maybe 35,000 across the whole of Europe.

Three years later in July 2006 it was the turn of Wisley in Surrey to see the temperature hit a sizzling 36.5C on the 19th of the month, the highest July temperature ever recorded. In Scotland high pressure also ensured that most of the month was similarly blessed with hot sunshine although it was cool and wet at times in the second week. Remarkably throughout the month the temperature still reached 25C somewhere in Scotland on twenty occasions and that included fifteen consecutive days by the third and fourth weeks, Prestwick Airport hitting the dizzy heights of 31C on the 19th.

What we faced in August 2003 and again in July 2006 is still seen as a taste of things to come during the next century, although subsequent poor summers in 2007, 2008 and 2009 have proved otherwise for the time being.

The summer flooding of 2007 stands alone as an event which could not have been foreseen, nor can it be regarded as a product of climate change at this stage, but it is worthy of special mention. Although the flooding was clearly exceptional it was to my mind a typical British summer of old and a reminder of why Mediterranean holidays first became so attractive to us more than forty years ago. While we were being drenched a heatwave brought temperatures of 40C or more into eastern Europe from Asia Minor. To some people the disparity may seem to indicate some sinister shift in our climate but it has often been that way. Very often, during the long summer months, temperatures are often much higher across central and southern Europe peaking as they do between 30C to 35C accompanied by blue skies and light winds. As the summer evolves, the jet stream normally moves the rain bands to the northwest of Scotland allowing the Azores high pressure to extend across Britain from the southwest. In 2007 that modifying weather pattern failed to materialise. History shows that 1912 was also an atrocious summer with torrential downpours and severe flooding which probably exceeded that of 2007. It must have seemed particularly bad at the time because 1911 had been such an exceptionally good summer.

So our most recent summer flooding is by no means unprecedented and as we all enjoyed a fantastic April there was already a feeling that the weather had peaked too soon that year. The early days of May were still dominated by high pressure centred over northern areas but by the second week there were ominous signs as rain systems broke through from the Atlantic. Heavy downpours began to develop and 13th May proved to be exceptionally wet across England and Wales as an intense low-pressure system moved in from the southwest. Some places had as much as 40mm to 50mm of rain in just twenty-four hours but after a few days of hotter weather – the Chelsea Flower Show enjoyed glorious sunshine – there was to be further torrential rain by the late Bank Holiday weekend. No play was possible on the Sunday of the Test Match at Headingley between England and the West Indies, although the heaviest rain was again across central and south-eastern areas of England. By Monday, London was cold and very wet with temperatures of just 8C compared to 25C just four days earlier.

The return of the heavy rain left many places in the south with their wettest May on record as thundery showers followed the more persistent rain into Scotland. June began on a brighter note as the Azores anticyclone linked with high pressure over Scandinavia to bring drier, warmer conditions to the worst hit areas the previous month. All went well until history repeated itself on 12th June as the high level jet stream spawned new intense depressions to the southwest of Ireland.

For most of the following two weeks it seemed to rain incessantly, although in reality there were a few brighter interludes. Only the far north of Scotland escaped the clutches of the deluge with the bulk of the country marooned and powerless in the pluvial war zone. Flooding became increasingly severe and widespread, although the western Midlands and South Yorkshire were now suffering most. In Sheffield 88mm of rain fell on the 14th. The ground everywhere was becoming saturated as wave upon wave of 'tropical' rain engulfed many areas.

By the middle of the fourth week, rivers had burst their banks and thousands had been forced out of their homes, not due to return for many months or even a year as properties were repaired. Hull and Doncaster were amongst the worst affected initially but as the heavy rain fell across Wales and the catchment area of the River Severn so its flood plain was becoming increasingly vulnerable. Following the wet weather of May it would be the wettest June for England and Wales since 1860. High pressure to the north was responsible for Shetland having only tiny amounts of rain but the same high pressure was acting as a block to rain bands heading north across remaining areas. As they hit the barrier some of them then moved south again enhancing rain amounts.

Locally in Lincolnshire the worst flooding occurred around 27th June with Louth and Horncastle residents now used to the regular sounding of the flood sirens. The River Till burst its banks onto adjacent fields within half a mile of our house and just a few miles northwest of Lincoln. Thoughts were turning to summer holidays as July arrived with the belief that the weather had to get better – it simply couldn't get any worse! But an improvement in the weather was not forthcoming and it did get worse – much worse. July would turn out to be the wettest since 1936 and the dullest for nine years. The pattern of the two previous wet months continued as the jet stream remained abnormally far south leaving us at the mercy of warm, moist air and intense depressions developing to the southwest. Occasional breaks in the rain and brightening skies indicated the potential of the warm air to lift temperatures significantly and the London area occasionally reached 25C in mid-month. However, July was dominated by heavy showers, a few tornadoes and lengthy periods of rain which reached biblical proportions. River levels remained high in many areas and flood plains were commonly under water adding to the escalating numbers of people forced to leave their homes. The most severe and

Bad business for car sales at Toll Bar – the community on the outskirts of Doncaster devastated by floods during the wettest July since 1936. (YPN)

Fury at what was seen as official indifference led the Prince of Wales to make a well-publicised visit to Toll Bar on 4 July 2007. He understandably looks a touch anxious! (YPN)

**The River Severn spectacularly burst its banks in July 2007, creating major media interest as communities like Tewkesbury were effectively cut off from the outside world. The twelfth-century abbey was only just clear of the surrounding floodwater.**

prolonged rain was probably that of the 20th when additional totals in excess of 50mm fell throughout southern Britain leading to extensive flooding across Berkshire. Pershore College in Worcestershire made the record books when a staggering 157mm was recorded in forty-eight hours – it just never stopped raining and most of it was heavy from a sky as black as ink! Inevitably the floodwater from the River Severn, Avon and Thames was now extensive across large parts of the south and west Midlands with Gloucestershire, Worcestershire and Oxfordshire particularly hard hit.

The flooding in the Tewkesbury and Gloucester area became the major news story as the Castle Mead electricity substation was shut down on 23rd July and water treatment plants failed leaving 350,000 without mains water until 7th August. Locals are used to the River Severn bursting its banks, despite improvements to the flood defences over many decades, but this was said to be the worst flooding in the area since the rapid thaw of snow after the severe winter of 1947.

In all thirteen people died during the summer floods of 2007 and 48,000 households were flooded.But it was not the wettest summer ever, 1912 holds that record, and it was probably not a result of global warming. It could happen again!

# THE GREAT STORM OF 1987

The Great Storm on the night of Thursday 15th October 1987 is indelibly stamped on the mind of every weather presenter and forecaster of the time. In the media Michael Fish became the scapegoat because of his story concerning a woman phoning the BBC earlier that day asking if there was a hurricane on the way. "Well I can assure you, there isn't", Michael confidently replied. And he was right!

But the story of the Great Storm goes back to the previous Sunday lunchtime when I was presenting the forecast on BBC1 for the week ahead, a forecast traditionally aimed at farmers and growers because of its position in the schedule within what is now 'Countryfile'. Towards the end of the broadcast I specifically referred to the "possibility of some very strong winds reaching Britain at the end of the week". This is a broadcast which to my knowledge still exists in the BBC Weather Centre archives but is never shown on any documentary programme about the Great Storm. From a media viewpoint it is too close to what actually happened and perhaps is regarded as uninteresting.

I was at pains to point out that something very nasty was brewing in the Atlantic, although at that stage it was still a figment of the computer's imagination, such are the complexities and fluctuations of any weather forecast at that range. Of course there was nothing unusual about referring to a possible spell of severe weather at that time of the year. After all it is the traditional period for equinoctial gales when the North Atlantic spawns several of these events, aided by the huge amounts of energy freely available in the warm waters of the Gulf Stream emanating from Central America where true hurricanes can wreak havoc.

So I went home on the Sunday lunchtime believing that subsequent computer analyses would reinforce the development of an area of low pressure as the week went on and it would become rather more than just another autumn gale. I have said many times that extreme weather is what my job is all about, easily outweighing the numerous dull days for which Britain is prone.

**The morning after! Two police officers patrol in London's Parliament Square amid debris stemming from the capital's worse storm in living memory. (AP/Press Association)**

On the Monday and Tuesday I was back at the London Weather Centre on radio duties, not due on television again until Friday morning when I was to stand-in for Francis Wilson on BBC Breakfast. A ten-hour radio shift would run from 8am to 6pm to include a host of BBC local radio scripts and broadcasts together with the Radio 4 12.55 and 18.55 broadcasts. New computer runs on the Monday and Tuesday continued to show some vigorous developments and a deepening low pressure system for the latter part of the week, as previously. Significantly though the expected passage of the severe weather was now slightly further south leaving France at the mercy of the storm.

These were the first signs that Britain was not likely to experience the strong winds I had hinted at on the Sunday, with the exception of the Channel coast where gales could still bring problems. The new scenario was disappointing, not because I wanted to see a major storm rip across the British Isles but the fact that my earlier warning was now looking unnecessary and computer predictions were moving the path of the storm southwards. My colleagues accepted the changes, as you have to.

So it was just one of those things. A change in emphasis which would leave Britain in relatively benign conditions at the end of the week but the realisation that just one grid point (or sixty miles) error from the computer could still bring trouble.

For any weather presenter a couple of days off should mean just that, so on Wednesday and Thursday I aimed to enjoy the break. But this was different, although I didn't know it at the time. Of course no-one knew what was about to unfold, not even the powerful weather computer at Bracknell!

I went to one of my local pubs on Thursday lunchtime for a couple of beers. Even then I sensed a feeling of 'calm before the storm'. Hindsight is a wonderful thing, as we often hear, but there was an eerie silence in Hertfordshire that afternoon. I missed the infamous Mike Fish broadcast but as I was due an early start the following morning I intended to watch the 21.30 BBC1 presentation. As far as I was concerned the story remained the same and France would catch the storm. We wouldn't!

Bill Giles was the presenter that evening and he rightly continued to keep Britain away from the severest of the weather although it would get "breezy up the Channel" overnight. Remind me never to go sailing with Bill!

I went to bed that night with few concerns. My car to take me to the Lime Grove television studios was booked for some ridiculously early hour but otherwise all was well with the world.

What happened subsequently was nothing short of mayhem as the Great Storm moved further north than predicted, not by one computer grid point (say 60 miles) but almost three. I was disturbed by the strengthening wind overnight even though Hertfordshire was a little way from the worst of what was happening.

The temperature rose 9C in twenty minutes at Farnborough, Hampshire in the early hours of Friday morning as the centre of the low pressure made landfall in Devon, drawing the remnants of tropical air north from Biscay. As we travelled south along the A1 there was evidence of storm damage as trees littered the hard shoulder on the higher ground around Welwyn. Reaching Greater London between Mill Hill and Cricklewood it was still very windy but it was obvious that something much worse had recently happened as we were travelling with all the traffic lights out of operation. In fact there was no light anywhere in these north London suburbs, only complete darkness apart from occasional car headlights along the way.

The true realisation of exactly what had struck southern England just a couple of hours earlier was still unclear but as my car arrived at the studios it soon became obvious. There was a major power outage in the whole of west London, including Lime Grove studios. The production team and presenters from the BBC Breakfast News programme were standing outside, some of them shining torches. I was a marked man. It was my responsibility and what had I got to say.

For once it was comforting to tell them I was only the messenger. When the forecast goes well it is with a feeling of self-congratulation that you beat your chest and take the plaudits.

Jeremy Paxman was presenting the programme that morning and you don't mess with Jeremy when he is in a mood.

"I don't know what went wrong and I have no more information than you until I can ring the London Weather Centre," I exclaimed. Mobile phones didn't exist of course.

We had one hour to get on air but there was no power at Lime Grove so everyone made a quick dash on foot to TV Centre half a mile away. For some reason the only power available to get on air was from the Children's continuity studio, conveniently next door to our normal weather studio. In the words of Les Dawson, talking about the Nottingham studio where I made my debut seven years earlier, the continuity studio was no more than a "technological broom cupboard". It was tiny with barely room for two people to sit in front of the unmanned camera fixed in one corner.

As 7am arrived Nicholas Witchell began a special news programme of sorts until 'Breakfast Time' could obtain more information and somehow access pictures of what had happened overnight. There were no slick graphics to use as we were in temporary accommodation and for my part I was woefully lacking weather data. I managed a telephone call to the London Weather Centre where they must have been running on their stand-by generator for the past few hours. The information passed to me was scant but no-one was taking prisoners that morning. I had to say something! The centre of the storm would soon be in the North Sea, leaving England around the Wash. Winds had been gusting in excess of 100mph in the last few hours across southeast England, and there would be a slow improvement through the morning – and that was about it!

Ian McCaskill was the senior forecaster on duty and he was due to finish at 8am but the day shift never arrived – they couldn't get to work and Ian couldn't get home.

For my part I was called upon to join Nick Witchell at about 7.15am after he had valiantly tried to explain to the rest of the country, who still had power, that a 'hurricane' had swept across southern England during the night and there was a trail of destruction and devastation. Mike Smart had compiled what pieces of information he could about damage to property and communication links and now it was my turn to take his place in the studio alongside Nick. "Obviously weather information was very limited at this early stage and the worst is not over yet" was my first shot from the hip! I tried to explain what had been happening through the night, where the storm had come from and where it was now. All I really knew then was that central and northern Britain had seen nothing in comparison to the south where power lines were down and many of those affected by the devastation could not even see our programme going out.

During the next couple of hours some sense of normality slowly returned to the BBC as power was restored to another studio adjacent to my normal weather studio. It was often used by whispering Bob Harris presenting 'The Old Grey Whistle Test' among other things but that morning no-one was whispering; everyone wanted to be heard and I was no exception. Jeremy Paxman was called into action and the start of the post mortem began.

A normal schedule would mean five or six weather broadcasts sprinkled within 'Breakfast Time' but this was by no means a normal schedule; in fact there was no schedule at all as news was frequently being received during that Friday programme. Time and time again I was asked questions by Jeremy Paxman about the latest situation across the country but I could only wonder how many people were actually watching this broadcast in the light of the vast numbers of power lines strewn across the countryside. For the millions in the northern half of the UK where there had been no weather of any consequence there must have been a sense of bewilderment but at the same time the belief that all this fuss was being made just because it was affecting London.

Well let me say that this amazing storm did impact on the lives of many more millions other than those living and working in the south of England, not least because of the travel disruption into and out of the capital from places as far away as Scotland and Northern Ireland.

Shortly after 1pm I was relieved of my duties as the 'morning' presenter, a shift which normally ended at 9am consisting of five broadcasts had expanded to almost twenty. Over at the London Weather Centre, Ian

McAskill was still on duty, now having to face severe criticism from Michael Buerk on the lunchtime news. No punches were pulled by an angry Buerk who went straight for the jugular with Ian looking as though he had been up all night – which he had of course.

"Well a fat lot of good you lot were last night," was the abiding memory of that interview. Poor Ian could only duck and weave like a drunken man in the headlights of a raging juggernaut. He had no answers but probably achieved the sympathy vote.

I can only look back after all the intervening years and think that Jeremy Paxman must have been a 'pussycat' in those days and he had given me an easy ride.

A story lost in the mists of time is that just two days later on the Sunday afternoon the three of us – Michael Fish, Ian McAskill and myself – were to take part in a quiz programme called 'Masterteam'. This was to be recorded in Central London hosted by Angela Rippon between teams from around the country representing different occupations.

All very straightforward under normal circumstances but the 48 hours between Friday and Sunday afternoon were unique in the world of television weather presenters. We were all culpable after the events of Thursday night's devastation and the press would not let go. I had received an easy ride with just one visit from the press but poor old Michael and Ian were targeted for different reasons: Michael for his story about the woman from Wales; and Ian for the criticism on the lunchtime news as he had been the senior forecaster on duty that night.

As a consequence the recording of a quiz programme was more of distraction than the usual feeling of anticipation as we were all very tired. During the preparations for the programme the production team suggested to us that the first round should be fairly straightforward, the questions would not be too taxing and given the 'rub of the green' we should comfortably beat our opposition. Just for the record we were up against a team of metrologists, experts in weights and measures.

Angela Rippon began the questioning and it soon became apparent that the opposing team across the studio floor would 'measure up' far better than our team of experienced broadcasters. It would turn out to be an embarrassing defeat as general knowledge questions were met with few correct answers. Rather than enjoying the experience we all wanted the earth to open up beneath our chairs and put us out of our misery. After an hour or so it was all over and the humiliation was complete, but it was all due to one event two days earlier, the like of which had not been seen in our lifetimes.

As all the statistics of what has become known as the 'Great Storm of October 1987' were collated at the Met Office it became clear just how rare and severe it had been. Researchers gathered every piece of satellite, surface, radar and upper air information and ran the computer with

**The 'Tree of Heaven', one of the jewels of the Royal Botanic Gardens at Kew, fell onto King William's Temple at the height of the Great Storm of October 1987.**

staggering results. Taking the area of maximum destruction as being southeast of a line from Southampton to Great Yarmouth, including London, mean wind speeds and maximum gusts in this area were considered to recur no more than once every two hundred years.

The Great Storm was not technically a hurricane as it didn't originate in the tropics nor did the mean wind speed reach 75mph for at least ten minutes. But that mattered not one jot to the press or to the millions who had faced unprecedented disruption that Friday morning. Nineteen people lost their lives, fifteen million trees crashed to the ground and the cost to property and insurance claims ran into billions.

As the faces of the Met Office, the television broadcasters were seen as scapegoats for the whole sorry incident and our performance in front of the press in the first twenty-four hours could have been better. Rightly it was decided that we should all attend a half-day course hosted by the late BBC journalist Bernard Falk using his own media company. The experience proved invaluable as we were all given sound instruction and then individually questioned by Bernard on hypothetical situations of a serious nature. All the interviews were recorded for review purposes.

The bottom line was that as trained broadcasters we should not be riled by the questioning and should maintain our cool and discipline at all times, thereby remaining authoritative.Perhaps we should also have been advised not to attempt a quiz programme when the full media spotlight was still on us!

At the Met Office it was also a time of great stress as questions were being asked in the House of Commons. Over twenty years ago there certainly was not the computing power available that there is now, nor was the quality and depth of raw data so great, but it was a public relations disaster. Journalists naturally accused the Met Office of failing to forecast the storm correctly and it didn't matter to them that the threat of a severe storm had first been highlighted on the previous Sunday morning.

An internal enquiry was also undertaken and in time, following input from independent assessors, several recommendations were accepted. Importantly the observational coverage of the atmosphere to the south and west of the British Isles had to be improved. The developing storm of 1987 had 'slipped through the net' of the observational network at the time and that could not be allowed to happen again.

Kent and Surrey were perhaps the worst hit counties suffering massive damage and prolonged power cuts. One year on from the storm I was commissioned by the educational department at the BBC to narrate a couple of schools programmes, dealing with the weather for the 'Landmarks' series. One was to specifically highlight the repercussions of weather disasters on families around the world, be it droughts in Africa or mass flooding from cyclones in Bangladesh. It was fitting that a once in two hundred year event likened to a hurricane and striking southeast England should also be included.

At Sevenoaks in Kent the seven oak trees on the green had stood since 1902 but in a few minutes on that fateful night the seven oaks had been reduced to one oak. The night was commemorated exactly one year later – 16th October 1988 – when a time capsule was buried in the Vine Gardens close to the cricket field. Memories of the night from the young and old were placed in the lead-lined casket for future generations to uncover in the fullness of time. A special edition of the local newspaper at the time, photographs, a history of the local school, cassettes with personal memories and a candle were just a few of the contents.

Elsewhere across southern Britain millions of people had a story about that unforgettable weather event. One harrowing ordeal involved the train carrying dozens of holidaymakers from London to Gatwick. As the train pulled out of Victoria at 1am that Friday morning all they had on their minds would be the week or two of relaxation in sunnier, far-flung places around the world. At that time the wind was freshening but by no means exceptional for an October night.

Suddenly a terrific grinding noise caused panic among the passengers and as the train entered Merstham tunnel the noise increased and then the lights went out as the train shuddered to a halt. In the darkness the news got through from the front that the train had hit a tree which had ploughed into the tunnel and was wrapped under the carriages.

There was complete darkness apart from the weak glow from cigarette lighters and as time ticked by confusion turned to fear. The noise of the wind at the mouth of the tunnel was likened to a scream and it was increasingly obvious that the situation could be serious. Eventually a diesel train from Redhill came down the track and pushed the stricken train through the tunnel to safety so all was well .... Or was it?

What happened next you couldn't make up. Incredibly as the train came out of the tunnel it hit another tree which had fallen across the cutting onto the tracks. The winds were now howling and the scene could have been straight from a horror set at Pinewood Studios. A rail guard and two off-duty police officers told everyone to abandon the train for safety reasons.

All the passengers, many with luggage for their flights from Gatwick, were told to leave the train empty-handed and walk down the track to Merstham Station for protection. So they struggled into the teeth of the now fierce wind along the railway line for almost half a mile as trees and branches fell around them.The roar of the wind through the trees was punctuated by the sound of breaking glass, the passengers eventually seeking some solace under the pedestrian bridge close to the station where they huddled together and prayed, probably hoping this was just a dream.

Further disappointments followed as their rescue train was also stopped by more storm damage near Earlswood, so it had to reverse back to Redhill. At last they could seek refuge in the Post Office canteen but then a power cut meant they had to leave the building for security reasons and return to the stricken train. All around them windows and signs were being smashed but some made the effort to try and sleep until eventually at 10am coaches were employed to take them to Gatwick, avoiding fallen branches and debris along the way.

For many on the 1am from Victoria that night it was the end of a nightmare. They might have missed their scheduled flight but no holiday can ever have been more welcome when they did eventually get away.

Ten years earlier I had been living in the area enjoying five very happy seasons playing cricket at Farnham, Surrey from 1975 to 1979. I could never have imagined this quintessentially English landscape would be so severely damaged by a storm of unprecedented force in 1987. Here was an area of great charm which I instantly fell in love with at the time, although it was too expensive for me to afford a property.

The small market town and surrounding villages were situated right in the path of the trail of destruction that night, though I hope none of my old friends living there would hold me personally responsible. Candlelit dinners became a way of life as there was no power for over a week, those lucky enough to have a log fire being in great demand no doubt. Immediately after the passage of the storm all roads out of Farnham were blocked, many by fallen trees, and in Farnham Park adjacent to my old cricket ground only one quarter of the mature trees survived the ferocity of those winds.

That month had been unusually wet in the southeast of England with locally 150mm (6 inches) of rain being recorded in parts of Surrey. As this was already twice the average for a typical October, often seen as a wet month anyway, the ground was becoming saturated and unstable as the jet stream continued to sweep active bands of weather further south than normal.

It turned out to be the wettest October of the century for many and with it came perhaps the first rumblings of 'climate change' as 1987 had experienced more severe or dramatic fluctuations in weather than normal.

Back in January, Siberian conditions swept westwards out of northern Europe bringing plunging temperatures and then blizzards by the second week. Transport was brought to a halt, schools and offices remained closed and millions stayed at home. April then turned out to be the warmest since 1961 followed by a cool, wet June. Late summer and early autumn were reasonable before several incidents of flooding occurred in early October prior to that fateful night.

As bad as it was by the morning of 16th October it has to be remembered that because the storm came overnight not many people were out at the time. The fact that nineteen died was bad enough but if we had been faced with the same developing weather situation even six hours later as people left for work there would have been an even greater crisis.

# THAT SONG IN 1988

The band was a 'Tribe of Toffs' and the novelty hit 'John Kettley is a Weatherman' made number twenty-one as the highest climber in the Christmas charts of 1988.

A civil servant working for a branch of the Ministry of Defence would normally shy away from public attention but the opportunities bestowed on them as a weather presenter make them public property. On the other hand for a band to write a song about them is surreal, unique and totally unexpected. Stephen Cousins, Andy Stephenson, Phil Rogers and Michael Haggerton were the likely lads who helped to put this weatherman 'on the map'. So how did it all come about, and why me?

The first I was to hear of it was in February 1988 when a small audiocassette arrived in my mail at Television Centre. An accompanying letter from the band wished me well in my broadcasting career but the main thrust of the letter was to explain that they had written a song about me and hoped I liked it. In May that year they had been invited to play at a concert in London and they would be thrilled if I could appear at the venue and introduce them to the stage. Let me say that the recording left a lot to be desired but I suppose these were the days long before digital quality. As for the song it was 'of its time' with topical names and blessed with a catchy chorus:

> *John Kettley is a weatherman, a weatherman, a weatherman.*
> *John Kettley is a weatherman, and so is Michael Fish.*

This was the culmination of hours of fine-tuning in Stephen's bedroom studio at his parents home in Sunderland. The boys were in the sixth form at the time but had been writing and performing since 1986. Surely life in Sunderland couldn't be so dull that this humble weather presenter was an inspiration for the track which would ultimately catapult them to stardom, albeit briefly.

To be absolutely honest I was not totally convinced about the authenticity of the letter and the song. Maybe it was some kind of hoax or a wind-up. In the end I turned down the chance to meet them and introduce them to their adoring fans at the London concert. For me it was still a bit of a joke and I replied apologising that I wouldn't be available as I was playing cricket. Not totally unexpected as I was still playing village cricket fairly regularly, and

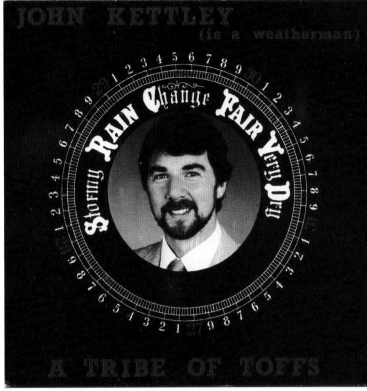

**Sleeve of the 1988 Tribe of Toffs hit 'single' 'John Kettley (is a Weatherman)', which became the highest climber in that year's Christmas charts and went on to sell over 100,000 copies.**

given the choice I would rather be strutting my stuff in the Hertfordshire countryside than at a noisy venue in London.

That would be the end of it as far as I was concerned but I wished them well.

So the Tribe of Toffs May concert came and went and life continued as normal – or so it seemed. Apparently Children's BBC also received an audio copy of the song and they had an idea for the summer kids morning programme 'But First This'. It was now July and producer Paul Smith decided to call Stephen Cousins to ask if he would like to bring the band down to London and record 'John Kettley is a Weatherman' at their Maida Vale studios.

The band couldn't believe their luck. Just eighteen years old they were now being offered the chance to record their novelty song for BBC television and would be given accommodation at the Kensington Hilton, just a short distance away from Shepherd's Bush, for free!

By late July I was also contacted by Paul Smith. Out of the blue, and some six months after the first contact, I was to meet the band face to face – but it would be no ordinary meeting. Paul Smith's idea was to promote the song by producing a pop video, normal procedure even more than twenty years ago. So why not bring the weatherman and the band together. His creative talent conceived a scenario whereby I would be presenting a weather forecast in front of the maps and the four members of the band would break through the screen behind. Suitably shocked I was carried away from my weather charts manacled in chains and masking tape – the first kidnapping of a weather presenter on British television.

At this point I should explain that the weather graphics were produced by a clever mix of back projection and colour separation overlay. The system required the screen to be flooded in blue light in order to produce the clear television image seen at home. For the purposes of the video the normally solid screen was replaced by a thinner plastic which allowed the boys to get from one side to the other.

On 9th August the video was made in Studio A at Television Centre, although we needed three attempts to get the best take. So the theme of the 'But First This' programme one August morning was the kidnapping of the weatherman and had anyone seen him? Would there be a ransom to get him back?

From the first airing on 'But First This', the boys performed at a few charity concerts and independently we did radio interviews to promote the song. *Radio Times* also ran a feature about the band and why they had been inspired to write the song. Several frustrating weeks passed and it was looking very unlikely that the video and the song would capture the imaginations of a record company. Eventually though a deal was signed in late October with Neat Records from Wallsend near Newcastle, although they were better known for their heavy metal artists. Bands called 'Snake' and 'Venom' were part of their portfolio and were big in France but the 'Tribe of Toffs' clearly bucked that particular trend. The song was finally released in early December and eventually reached number twenty-one in the Christmas charts selling over 100,000 copies along the way.

In early 2009 I caught up with Stephen Cousins in London to find out a little more about the history of the song and what experiences the Tribe of Toffs had enjoyed. Undoubtedly he was the driving force behind the band because all of their early recording had been done on a four- track cassette recorder in his tiny bedroom, no more than three metres by two with a sloping roof for the drummer to sit under.

The drummer, Phil Rogers, had the original idea of writing the song with some clever lyrics but Stephen had been instrumental in composing the music. Originally consisting of just two verses in late 1987 it was lengthened to incorporate the names of other people in show business or the public eye at the time.

I asked Stephen about the 'gig' they wanted me to appear at in May. Surprisingly, and something I had never realised, the four of them were all members of a Methodist youth club and they were due to appear at Westminster Central Hall at an annual event for the National Association of Youth Clubs. Stephen himself was the son of a Methodist minister and Sunderland was one of the positions his father had taken during a nomadic career.

Around 3,000 teenagers turned up and the Tribe of Toffs went down a storm, according to Stephen, largely the result of playing the one song 'John Kettley is a Weatherman'. They loved the chorus and it got everybody dancing so it was at that point the band thought they could take it further. The band got a 'buzz' from seeing the reaction of their largest audience and Stephen told me that it was obvious everyone could associate with this already well-known weather presenter. He believed that both the band and the teenagers thought of me as a person with a sense of fun, not least because of the sweaters I wore on daytime television and before 'Grandstand' on a Saturday afternoon!

Following the concert the boys completed their A Levels but there was a twist in the tale which could have scuppered their high profile television appearance later in the year.

Andy Stephenson left the band after we had produced the video for children's television as he wanted to pursue a career in the church. Already studying at Bible College, in October he was quickly drafted back into the band in time for the anticipated record release. On 6th December the band was booked to appear live on the Wogan show at the Shepherd's Bush Theatre. Initially the college refused permission for Andy to appear because their students had to be dedicated to religion and to their studies. It required several phone calls from BBC producers and from the record company to try to get the college in Sheffield to change their minds. Eventually they succumbed and for one night only Andy was granted permission to travel to London and perform with the 'Tribe of Toffs'. Although they were live on stage they did use a backing track but Terry Wogan was very kind when he met the band after the show, even remarking that he thought is was a great song and wished them the best of luck. The fairly elderly audience did not quite share his enthusiasm apparently.

I have never known to this day whether they also wanted me on the show but I have no knowledge if that was the case. Of course these were the days before mobile phones so I was out of range at Huntingdon races, oblivious to the whole episode at that point.

BBC Radio 1 had the sole rights to play the song in the week prior to its release but it was soon clear at the weekly playlist meetings that Simon Bates would be the man to plug the record.

Stephen was keen to point out to me in our discussion that Simon played the song twice in a row on his morning show on Friday 2nd December. He was in the bath at the time but after the second play he rang Radio 1 with a view to thanking him for promoting the record so much. Clearly it could have been a hoax call but presumably having already spoken to BBC people previously at the radio station they were convinced he was 'Bona fide'. Speaking together on the telephone and 'off-air' Simon said: "I think we've got a hit with JK so it would be great if you came on 'live' at about 12.15 to talk about all the stuff to do with the song". They went ahead and apparently the interview went very well which could only help to sell more copies once the record was in the shops.

I know now that Steve Wright didn't like the song. Shame on you Steve!

Of course every vinyl record in those distant days had a 'B' side and few people will remember that on the flip side of my record could be found 'Festive Frolics from Four Fellows', a catchy little title for the Christmas market.

Unknown to me at the time this was performed and recorded at the Radio Newcastle studios. John Ryan was one of their main presenters and had become good friends with Stephen. Clearly seeing an opportunity and having the connections in the area he was instrumental in sealing the deal with Neat Records and as part of the agreement he produced the flip side in his studio. To differentiate between the punk arm of Neat Records pop songs were released under the label 'Completely Different Records'.

Amazing what you learn many years later!

So everything was signed, sealed and delivered by early December 1988 and with the publicity wheels in motion it was onward and upward for the band. The high profile coverage from Simon Bates at Radio 1 and the live appearance on Wogan were invaluable. Stepping back from the front line I was being interviewed by local radio stations around the country and the band made a further television appearance on Daytime Live, including an interview with Alan Titchmarsh. According to Stephen the viewers would probably have learned more about Yucca plants than they would about the history of the song but maybe that didn't matter in the final analysis, and anyway Alan loved it.

The record apparently sold almost 110,000 copies, which to me is a staggering total and I still have two of the original six given to me by the record company. I tried to arrange a deal to receive royalties but negotiations proved fruitless to say the least so I was grateful to be the recipient of a handful of original vinyls.

To say the Tribe of Toffs were a 'one hit wonder' is clearly true because attempts at a follow-up record were doomed, perhaps for a very good reason as its title was 'Oh No, Terry Wogan's on TV (Again)'. When you consider our record was given a prime-time slot on his television show it was probably disingenuous to release a song with such a title about one year later. I understand that Mr Wogan was not that pleased and would not support the song!

So the Tribe of Toffs career was short-lived and although they continued to write and perform for a while, the band eventually went their separate ways. Stephen juggled a career as a psychiatric nurse with playing in another band called 'Ruth'. Michael became an IT consultant and Andy also went the way of the National Health Service working in the psychiatric unit of a hospital. As we spoke at the beginning of 2009, Stephen Cousins still has no idea what became of Phil beyond his course at Manchester University, but at least his own career has taken him back onto television with appearances on the Lottery and also 'Never Mind the Buzzcocks'.

Many people have asked me over the years what I thought about the song and did I mind? Did they need permission and how much money did I make?

There is an old saying "Any publicity is good publicity" and in this case that is true. No price can be placed on an episode as huge as this and the fact that people still refer to it constantly over twenty years later speaks for itself.

The boys never appeared on 'Top of the Pops' because in the week they were booked the song actually slipped four positions in the charts. It was devastating for the band but their loss turned out to be my personal gain as I was invited onto a 'Top of the Pops Special' to commemorate their own twenty-fifth birthday, the first broadcast having been on New Year's Day 1964.

My notoriety as a result of the song meant I would be mingling with all the disc jockeys and presenters of the programme from Jimmy Savile through to Mark Goodier. I enjoyed plenty of banter with Alan Freeman, Tony Blackburn and Paul Gambaccini, but when people ask me about my most embarrassing moments on television this show ranked quite highly.

The researchers asked me to 'blend into the background holding an umbrella' – presumably to emphasise the fact that I was a weatherman! Firstly it is impossible to blend into the background whilst holding an open umbrella in a studio for ninety minutes with live bands.

Secondly I was usually in the foreground grinning like a cat with his saucer of milk being jostled by the audience. What an experience but did I look stupid?

Twenty-one years later and it would be fantastic to get together with the 'Tribe of Toffs' again, maybe to re-record the song. Obviously I owe them plenty but this wasn't about financial gain but more ensuring the longevity of a humble civil servant doing a job he enjoyed most. And so did Michael Fish.

## 9.

# LONDON AND THE BBC WEATHER CENTRE

After five years 'learning the trade' in Nottingham I headed south to London on 29th July 1985 seeking the bright lights and national exposure. My first port of call was to be the London Weather Centre in High Holborn, then the home of Radio 4 weather bulletins, leaving all BBC national television broadcasts to be shared between the White City and Lime Grove studios in West London.

Two radio stalwarts at that time were Michael Shaw and the genial Irishman, Frank Greene, so it was a thrill to meet both on my first morning. I believe it was merely coincidence but they had arranged lunch with none other than Brian Redhead, anchorman on the 'Today' programme he shared with John Timpson. We enjoyed a convivial meal at a local Italian restaurant although it was more a question of listening to Brian rather than making too many contributions. At one point he asked me what kind of broadcasting experience I had to my name. More significantly he gave me just one piece of advice which I remember to this day: "Don't try to be too clever. I can tell you have a sense of humour but keep it under wraps until the listeners are comfortable with you. Gain their confidence."

Brian was a very experienced broadcaster with a point to make and I understood why he made the observation. It was a great pleasure to meet him and I was grateful for his comments and looked forward to working on the 'Today' programme in the future. Sadly we never actually met again face to face due to his untimely death in 1994; our brief conversations prior to my forecasts would be from separate studios.

After a few days settling in, undergoing some radio training and familiarising myself with other duties whilst on shift, it was off to BBC Television Centre. I had been chosen to replace Jim Bacon who had recently been promoted and as such had to leave for another position at the London Weather Centre. Everyone knew the system was wrong. Why should such a fine broadcaster be allowed to disappear from our screens on promotion, but Jim's loss was to be my gain.

**A posed 1985 press photograph intended to suggest that I was about to be 'licked into shape' prior to becoming a fully fledged member of the BBC weather team.**

My work would be split between the two centres in blocks of a few weeks. Despite the experience I had gained these were new challenges and clearly this was to be a steep learning curve. Bill Giles was the boss who had the task of training me for national television but as if that wasn't enough there was also a new computer graphics system to understand. Still in its infancy the system had several teething troubles which meant it occasionally went wrong, even during a live broadcast. For me that was probably the most nerve-wracking aspect because if you didn't have the graphics to show it suddenly became a radio broadcast on television – and there was nowhere to hide!

At the end of the first week of training I was far from comfortable with my surroundings and it must have come as quite a shock to Bill when I said, "If I am not feeling happy after my first month I would prefer to go back to Nottingham."

At the weekend I drove into the Derbyshire Hills to gather my thoughts. I didn't feel happy but I knew it was partly because I was 'computer illiterate'. My meteorology was fine and I was happy in front of a camera but my confidence was low because of the sometimes unreliable equipment I was working with.

After a few weeks I made my debut in late October or early November with the advice of David Stevens from Pebble Mill ringing in my ears: "When you first appear in someone's living room you are there without an invitation. They didn't ask you to turn up. Be friendly, do not upset them and above all do not patronise them."

I did my best without setting the world alight and I knew it would get easier.

In those early days the team consisted of Bill, Mike Fish and Ian McCaskill, who had also left his broadcasting duties at Central Television in Birmingham. The weather office was situated in the area mostly occupied by BBC Children's Department where the multi-talented Philip Schofield was an up-and-coming presenter at the time. We enjoyed plenty of banter and it was a joy to watch and learn from his relaxed style. Our office was bijou but at that time there was just one person on duty for ten or twelve hours so you needed to be happy in your own company. It was just a short walk down two flights of stairs to the presentation studios where the broadcast was produced.

With just four broadcasts a day it was a job to enjoy with plenty of time to study the weather across the rest of the country. Conferences were held with the senior forecaster at Headquarters to discuss the technical guidance issued with every computer run. It was the job of the presenter to turn that scientific information into a weather story which could be understood by the man or woman at the bus stop. It sounds easy but a complex story needs simplifying to make it comprehensible, and there lies the skill.

**Self-operating office weather studio at the TV Centre, 1986.**

The fact there was time in the day to keep in touch with the weather and to answer your mail was something that was compromised in later years with an increased work load. During the late 1980s there was more than enough time, indeed as a member of the BBC Club I could afford the luxury of a break in the bar immediately following the early evening broadcast – known as 'networking' these days. Many a drink was shared, sometimes with the great and the good along with directors and producers working behind the scenes on other programmes.

On Saturday afternoons it was a pleasure to meet Des Lynam in the 'Grandstand' studio, where we would occasionally exchange thoughts on the winner of the 2.00 at Newbury or what might happen later in the football matches.

This was a brilliant working environment and I was delighted to be asked to appear on light entertainment programmes during the first year or so. Now I was more comfortable 'with my feet under the table' and the prospect of working in other branches of television was a thrill.

'Blankety Blank', originally made famous by Sir Terry Wogan, was my first quiz programme in which I appeared alongside Ian and Mike. The Beverley Sisters provided more attractive contestants and Les Dawson was the new host of my first show in front of an audience. I have little recollection of how the programme went, although I have an everlasting image of a little man turning a wheel behind the set which brought the contestants round to face the audience! Les was also a perfect gentleman being very complimentary and charming in hospitality afterwards.

In November 1986 I again had good reason to be grateful to Jim Bacon for getting me a contract which would bring me vast experience during the next three years. Jim had been the resident weather presenter on the BBC2 programme 'Travel Show' which was normally shown 'live' from the Manchester studios. Following Jim's promotion the executive producer, Alan Dobson, needed a new presenter and Jim very kindly recommended me as the 'new kid on the block'. A Christmas Special covering winter holidays was required in mid-December so there was no time to waste. I met Alan in Manchester. We got on extremely well and agreed a contract for the programme.

This was a fantastic opportunity and again it was a new experience. A live BBC2 programme presented by Paul Heiney. I was delighted the way it went but more importantly so was Alan. During the following three summers I would be the resident weather presenter on thirty-nine live programmes, usually transmitted at 20.30 on Thursday evenings. We also recorded eighteen Travel Show Guides on November weekends which gave more in-depth coverage of particular destinations such as Disneyland Florida, Cyprus and the Canary Islands.

All the programmes were produced in Manchester which meant I had to take a day of annual leave but I could be back on shift for radio or television at any time on the Friday. Usually I travelled by train from Euston to Manchester staying overnight at the Midland Hotel, all expenses paid. Occasionally an early morning start on the Friday meant leaving on the 01.30 sleeper for London to be at work for 7am. A few times I even flew, either with Suckling Airlines from Cambridge Airport or British Airways from Heathrow.

All this was a far cry from making the tea at Manchester back in August 1970 but beginning as a scientific assistant made me appreciate the good times even more. Paul Heiney left the programme after my first year (purely coincidental) and he was replaced by the royal author Penny Junor. Matthew Collins was our roving reporter who was sent all over Europe every week and often arrived back in the studio for the programme with barely three hours to spare! Along with the dedicated team behind the scenes this was a slick and efficient television production and I will always be indebted to Alan Dobson for giving me such a fantastic opportunity.

In the meantime I was involved in a few other light entertainment programmes for BBC. In February 1987 I worked alongside a rising star by

**Summer 1987 edition of 'The Travel Show', which was transmitted 'live' and proved to be a fantastic opportunity.**

the name of Carol Vorderman. This was a science programme called 'Take Nobody's Word For It' which was recorded in the Bristol studios. Together with Professor Ian Fells it was my job to bring the outdoor weather into the studios by showing how to make a cloud or a rainbow among other things. There was also an interview where I explained how gathering weather data and feeding it into the Met Office computer eventually led to computer analyses which the weather presenter then had to translate into a meaningful forecast.

During the first weekend of August that year my athletic skills (or not) were put to the test in an 'It's a Knockout' style programme based at Butlins in Minehead. Along with the likes of Suzanne Dando, Ed Stewart, Bernie Clifton and Angie Bowie we all made fools of ourselves for a Trans World International production to be shown in the autumn. It was probably not my finest hour as we all attempted roller skating down the promenade, canoeing round the boating lake, recovering plastic fish from the indoor pool and the bucking bronco machine. There would have been far less memorable feats of endurance as well but thankfully the passage of time means I have completely forgotten what other silly things were involved.

Umbrellas to the fore! (Top) Guest speaker at Abbeygate College, Chester, in 1987. I was welcomed by prefect Sarah Ratcliff and faculty head Roger Slater. (Chester Chronicle)
(Bottom) The wedding of Lynn and I at Kirton Lindsey, Lincolnshire, on 12th September 1990. It was not raining – and the temperature was 21C!

**With Brian Johnston at the Old Trafford Test Match in 1989.**

In contrast one of my finest hours was as a lunchtime guest of Brian Johnston, 'Johnners', at the Old Trafford Test Match in 1989. This was part of a series of Saturday lunchtime interviews dating back to Trent Bridge in 1980 with the title 'View from the Boundary'. Those who listened to the great man during his broadcasting career will have appreciated his perceptive questioning and amiable nature. As an interviewee I have to say Brian was a genius because the conversation was exactly that, namely a cosy and informal discussion which could have been taking place in the lounge bar of your favourite pub.

As it was we were chatting on the third day of the fourth Test between England and Australia in front of a live radio audience. Sadly the Aussies went on to win the match and the series.

It was even more special because with me was my girlfriend Lynn, who became my wife just over a year later. She was the missing link to the 'card game of life' which I had mentioned during my final year at college back in 1976. Our paths had first crossed in the Common Room over a game of table football or snooker but we would not meet again until she wrote to me in late 1988 having recognised me on television. She had created a name for herself as Lynn N. Grundy, illustrator of children's books since her college days. Now we had met in the hope of making up the lost years and maybe starting a family, which we eventually did with the arrival of Charlie in 1992 and George in 1994.

Other great opportunities came along in 1990 appearing twice with Noel Edmonds within just a few weeks.

Apart from the undeserved honour of having a song written about me I was also the only weather presenter to be presented with a Gotcha Oscar. Maybe not a huge claim to fame, but it was perhaps testimony to my sense of fun and irreverent style of presentation on occasions.

I was called one afternoon by a researcher checking my availability the following day for a corporate video in Milton Keynes. It was very inconvenient so I declined, but surprisingly they called back (having contacted my Agent) within ten minutes doubling the fee! Now I was interested, especially when they promised it would be a couple of hours' work with a car to transport me there and back. So nothing suspicious so far, although the following morning I sat in the car for fully twenty minutes outside the studios at Milton Keynes. Eventually I went over to reception to find out what was happening.

"Sorry John, I'll show you to your dressing room where you can have a cup of coffee. The director will chat to you shortly about the shoot", said the researcher.

So the brief was to prepare a snappy corporate video showing salesmen how to sell a car to awkward customers. After make-up I was taken to the studio where a Ford Fiesta was flanked by a young man dressed in an orang-utan outfit – he was clearly my awkward customer although I still didn't see this as being particularly strange!

The shoot began badly as I tried to get to grips with the words on autocue whilst moving my subject around the car, highlighting the good points of the Fiesta. My orang-utan proved difficult to work with, claiming to have trained for three years at drama school just to appear with me in a costume.

I was taken back to the dressing room for the director to explain that if I concentrated on my lines and positioning then the 'customer' would not be a problem: "He was disappointed with his acting career so far and getting increasingly frustrated."

At this point, unknown to me, Noel Edmonds was replacing the actor so back in the studio there had been a switch inside the costume. Now he was much more animated and rehearsals progressed apace, although we were both locked inside the car for a while so we had to escape through the sunroof, just to add to the drama. Still I suspected nothing. After all the money was quite good and "I would be home by mid-afternoon".

Eventually we stood at the back of the car to consider the boot space when all was revealed. I suspect it was getting intolerably hot in the costume so Noel had no option but to bring proceedings to an end. My reaction was understandably one of shock but also relief because I couldn't imagine that this corporate video was ever going to work. "I didn't know the series was still going" was my instinctive reaction, referring to Noel's 'Saturday Show' for which I was now booked.

This was a top programme on the television ratings and despite not acquiring permission from the Weather Centre and the Met Office to do the video I was delighted to be involved, if slightly embarrassed! Interestingly, Eamonn Holmes was caught with the same prank during the series and I have always wondered if he was in the studio before me, hence the delay.

A few weeks later I appeared on 'Telly Addicts' at Pebble Mill which was an evening recording followed by an overnight stay at the Holiday Inn in Birmingham before heading back to London the following morning. Noel Edmonds was again the host in front of a live audience, a situation I was now becoming familiar with and enjoying.

Also on the team of so-called celebrities was Ray Allan (without Lord Charles), Barbara Windsor and the late Jill Dando. Our opposition was the winning family from that year's series and they were expected to win comfortably. The quiz was great fun and again a privilege to have been approached, and this time I did ask permission. Amazingly we also won.

By 1991 I had been promoted again to Senior Scientific Officer and was now working totally from the newly formed BBC Weather Centre. There was no longer a requirement to share working time at the London Weather Centre but there was still a need for shift work and that included working nights.

As other programme opportunities came along I jumped at the chance and it was on one of these occasions I first met Dick Francis. The ITV production 'Through the Keyhole' approached me about filming our house close to Hitchin, Hertforshire. It was already our pride and joy and quintessentially English with a thatched roof and a small country garden. Without hesitation we accepted the contract in 1993 so it was an open invitation to Lloyd Grossman and the crew to film various rooms in the house, hopefully making complimentary comments as they did so. Our first boy, Charlie, was just a toddler at the time but they included his bedroom decorated with Lynn's work to highlight the fact that "someone in this house must be very creative" – clearly a comment directed at Lynn and not me with my scientific background. Filming lasted over six hours to be followed by the studio appearance with Sir David Frost at the Yorkshire Television studios in Leeds a few weeks later.

Each show highlighted two houses but four programmes were being made on that one day alone, some in the afternoon and some in the evening. In hospitality beforehand I was introduced to the famous Dick Francis along with his wife, Mary, and son Felix. Their property in the Cayman Islands served as a brilliant contrast with everything our home could offer but of

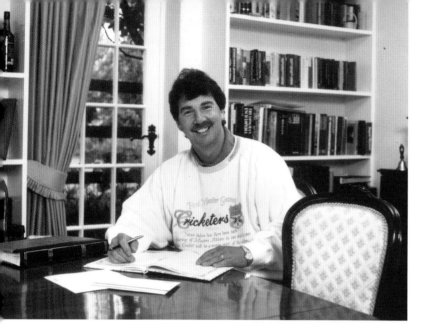

**In the study at our first home in Hertfordshire, which in 1993 featured in the ITV production 'Through the Keyhole'.**

course the programme researchers would have no idea that they were bringing together two professionals with a common link, namely horse racing. Dick, the former jockey, was now one of the greatest thriller writers in the world with a strong racing connection. As for me, well I was just a racing enthusiast but with an ancestor who was also an apprentice jockey.

We chatted at length before and after our programme about horse racing and also the weather. Dick was particularly interested in how weather forecasts were produced and confirming that I was actually a trained meteorologist. Felix now also managed his father's affairs having made a career change from being a physics teacher for seventeen years until 1991.

In the words of Sir David: "And now it's time to meet our panel....and what a panel they are." In this case Liz Hobbs, Nina Myskow and Chris Tarrant, who correctly guessed that the thatched cottage was indeed owned by a weather presenter with a strong Yorkshire connection.

Following an enjoyable day I returned south, never believing that a few years down the line I would become a consultant for a Dick Francis novel.

Back at the BBC we were now in larger premises with responsibility for all the national radio and television output along with satellite broadcasts into Europe. No longer just the four broadcasts per day but an increasing portfolio requiring more staff, more shifts and a greater commitment to our expanding weather graphics.

As the years went by the Weather Centre continued to expand with further new bespoke offices and two purpose-built studios operating simultaneously. The system we used for showing the weather graphics on television was known as colour separation overlay. There was a direct link between the BBC computers and the mainframe Met Office computer which would send the latest model information twice a day based on midnight and midday observations. All symbol charts could be edited to suit the way you wanted to show the graphics and present the forecast. It was your choice, and as the meteorologist on duty you edited the charts and presented the forecast in the way you believed was most appropriate. The autocue head attached to the weather camera showed the presenter exactly what was seen at home; in other words you saw yourself in vision at the same time. The perspex screen behind was flooded in bright light so it was a requirement of the system that when you pointed at the charts you were getting your directions from the clear picture in the autocue. A digital countdown clock was also visible with further instructions from the presentation gallery in your earpiece.

This was a quantum leap from the magnetic symbols I started with in Nottingham and even my early days in London when a floor manager would stand with his finger on an analogue clock to the side of the camera. So no earpiece, and just occasionally no floor manager either – that really was 'seat of the pants' broadcasting!

A lot of people are unaware of the fact that nothing was scripted. It was neither practical nor convenient as the weather bulletin was always seen as a flexible duration depending on 'breaking news'. "Far better to make it up as you went along," I always said.

By September 1996 the latest and most innovative changes to the Weather Centre saw the introduction of the thrice-weekly 'Weather Show' as well as the regular forty-eight television broadcasts a day and seven for radio. By now there was a team of fourteen presenters but improved technology also meant that Suzanne Charlton no longer needed to stand on a box. Instead the self-adjusting camera and personally-adapted lighting meant everything was automatic at the flick of a switch.

Inevitably staff numbers continued to increase and privately I began to wonder how much more we could take because the extra broadcasts into BBC World and News24 meant we were now operating round the clock. I would be the first to admit that the facilities were excellent but we were now more isolated within the BBC because all the broadcasts were produced and directed 'within our own four walls'. There was now little contact with other people around the building and we were becoming a 'sausage factory' for weather forecasts, without a sole.

My management responsibilities had increased, which was fine, but inwardly I was becoming a little disheartened with the isolation. Whenever the opportunity came along to appear on other programmes I would jump at the chance. Obviously competition was now more fierce with a greater number of presenters to choose from but I had the advantage of being there at the best time and I was now one of the most well known weather presenters in the country.

Bill would be retiring in early 2000 as the Met Office's boss of the BBC

Weather Centre. His counterpart at the BBC was the editor, John Teather, and by early 1997 he was making enquiries as to where I saw my future role within the Centre. At that stage it was perfectly reasonable to think of me as Bill's successor.

John was angling for my answer but for me it would not be the right move. By this time Charlie was four and our second son George was two. Above all I wanted to enjoy my children during their early years and to accept Bill's position would require trips abroad and a large reduction in my broadcasting duties in order to do his management position justice. On a positive note there would be less shift work but the decision was an easy one to make. I wanted to continue doing the job I enjoyed.

The untimely death of Diana, Princess of Wales, saddened millions around the world and for me it was also a pivotal moment in my time as a regular television weather presenter. It was Sunday 31st August 1997 and I was on duty that morning preparing for the lunchtime broadcast, which traditionally looked at the weather for the whole of the coming week.

That morning was unique. I prepared my charts as I always did but at the same time watched the television pictures as the shock news was unfolding hour by hour. For six hours this was a roller coaster of emotions which I sensed was beginning to affect me inside, even though I had never particularly been a supporter of the Princess, although I appreciated the charity work that she did at home and around the world. This was rolling news as we had never seen it before and the usual schedule was scrapped in order to cover the tragedy.

I was happy with my own story of how the weather would unfold but all that seemed trivial compared with the country losing an icon, even a treasure. By midday I was wondering if the weather would also succumb and my services would not be required. My time would not have been wasted as the afternoon shift could use all my prepared charts for later bulletins.

I was given the message from the presentation editor that they would need a two or three minute forecast in the usual slot, just before one o'clock. My reaction was almost disbelief because by this point I was emotionally 'drained' and I thought it frankly inappropriate to appear with a weather forecast after over six hours of constant news about the fatal crash in Paris.

When I eventually appeared I felt as I had never done for a broadcast before. Not shaking or nervous but totally overwhelmed. Surely a forecast was the last thing viewers wanted to see. To my mind what I was about to convey was trivial in comparison. I got through it but still felt surplus to requirements that morning.

By the end of the following year fresh horizons had brought renewed enthusiasm to my working life with two exciting ventures, both on a 'freelance' basis outside the BBC Weather Centre. They would prove to be key to my final decision to leave behind the work I had enjoyed since 1985.

1999 was the year of the Cricket World Cup and the vast majority of the forty-two matches would be played in England for the first time for sixteen years in a spectacular 'Carnival of Cricket'. A press release just before Christmas 1998 read as follows:

"Through the medium of television millions of cricket lovers around the world will watch the excitement and drama unfold. Here at the England and Wales Cricket Board, we will use that excitement and the colour and vibrancy of the World Cup as a catalyst to help expand and develop the sport into the new millennium. In an effort to extend the carnival beyond the field of play, we will encourage wide community involvement through culture, music, food and other events for all the family – particularly the young."

A letter to me from the Event Manager, Michael Browning, read: "As a cricket lover we are sure you would support these aspirations and it is with considerable pleasure we invite you to become an Ambassador for the event."

Clearly my passion for the game over many years was known to those in authority; after all I mentioned the weather for cricket matches in many of my broadcasts. I had also been a regular visitor to Test matches courtesy of Cornhill Insurance, for twenty-two years the official sponsors of Test cricket in England. My association with them went back to my time in Nottingham when I first met Bob Taylor, former England wicketkeeper and now a Cornhill representative. Since then I had also played a host of charity cricket matches, including many for the Lord's Taverners.

I accepted the role without hesitation, together with all the photo opportunities, dinners and cricket matches that went with it.

On 24th March 1999 I was invited to the indoor nets at Lord's along with other ambassadors and celebrities to meet the England players and enjoy 'a day in the nets'. As usual my shift work meant I arrived a couple of hours late but it was indeed a fantastic experience, although I declined the invitation to bat against Darren Gough, preferring instead to bowl for forty-five minutes at Graeme Hick – without success!

On Thursday 13th May I attended the NatWest Eve of Competition Dinner at the Royal Lancaster Hotel, Hyde Park, sharing the company of Ted Dexter, Christopher Martin-Jenkins and Sir Vivian Richards amongst many others. I was beginning to think I could get used to this.

It was patently obvious that the weather was likely to play a major part in this 'Carnival of Cricket' which was scheduled to continue until the final at Lord's on 20th June. As a sponsor, Vodafone were keen for me to provide weather forecasts for the major games using my voice on a dedicated 'forecast line'. Using equipment provided by Vodafone I updated the forecasts at home when required but it was a 'busman's holiday' for me with a handsome contract to supplement my income. Every game was covered by BBC or SKY and when I was on television or radio there was a constant reference to the cricket, an arrangement which the England and Wales Cricket Board happily supported.

Although May turned out to be the dullest for five years the rainfall was less than average and temperatures were continuing the trend of warmer weather. The temperature peaked at 27C in London on 16th June but it was a very chilly 10C in Dublin on 21st May for West Indies versus Bangladesh. Just one match failed to reach a conclusion, at Headingley on 6th & 7th June between New Zealand and Zimbabwe.

So to the final between Australia and Pakistan and both Lynn and I were invited for the 'best seats in the house'. There was a delayed start due to morning drizzle but then the weather improved. However, the match was an anti-climax as Pakistan fell way short of their best, so for the millions of cricket supporters around the world the 'Carnival of Cricket' ended disappointingly.

For me it was a success which I could build on one day – if I was brave enough!


To have met Dick Francis at the studios of Yorkshire Television in 1993 was a memory to cherish. I had read many of his more recent books following his first thriller as long ago as 1962. It was a complete surprise when the Francis family contacted me again in 1998 hoping I could help them with a new novel to be released the following September. Instead of a chat over a couple of drinks in a London hotel they had ideas far more familiar to me. They wanted to see my working environment and how the BBC Weather Centre operated on a daily basis.

What could I tell them about gathering weather information at one end of the operation to presenting a forecast for the media at the other end? We talked at length as Felix, Mary and Dick shadowed me throughout the afternoon, throwing questions at every opportunity to get a deeper insight into how a shift evolved.

The production team thought it would be a good idea to make a 'Weather Show' based on a 'hypothetical whodunit' which I could present. No problem there then, and Dick was happy to be involved as you would expect.

It was a beautiful early summer's day and the venue was Ascot Racecourse. I stood in the paddock area interviewing Dick about his career as a jockey and how he used his knowledge of the 'Sport of Kings' eventually to go from racing correspondent to crime writer. As I got to know more about him he learned more about me. We strolled along the track past the winning post, just imagining how different the place would be in a couple of weeks time for the Royal Ascot meeting. Dick also wondered what information I had as a weather forecaster that the Clerk of the Course would need. Could I also be part of an illegal betting scam with the information I had about possible heavy rain affecting the ground and favouring a particular horse – or not?

Opening ceremonies.
1. October 1998: 'Kingfisher
Way' – the Baldock to
Tempsford Riverside Walk.
(Biggleswade Chronicle)
2. January 1999: Ashwell's
new Post Office. (Milton
Keynes Photo Services)

From Ascot we travelled three miles west to Met Office Headquarters at Bracknell. We filmed in the Central Forecasting Office and also the computer room where the powerful Cray computer did the 'number crunching' to produce an accurate forecast. We both asked more questions with Mary and Felix attentive either side. By the end of the afternoon they had what they wanted for now and I had a 'Weather Show'.

As the weeks went by I chatted again with Dick and he built up his knowledge of my job and where a story could lead. Eventually the draft for his latest book was complete and it was my job to read through the whole novel to check for any meteorological errors. Time consuming it might have been, and I did change a few things, but from my point of view the latest Dick Francis novel Second Wind was ready to launch with me as chief consultant.

Lynn and I attended the official launch in early September at The Ritz in London and for one evening I was no longer John Kettley but Dr Perry Stuart, the main character in Second Wind and a senior forecaster at the BBC Weather Centre.

Since then I have kept in touch with the Francis family and have been a regular guest at the launch of the latest Dick Francis bestseller.

Behind the scenes there were other issues fuelling the unrest of the staff and it was no coincidence that the in-house monthly newsletter was issued under the banner 'Hidden Agenda'.

On 29th September an article in the Manchester Metro was typical of information appearing in the press by this time without foundation. I was now described as "the veteran weatherman who had twice tried to quit in rows about his future". Apparently my first letter of resignation followed reports that I had been unhappy about being told to cut my hair rather than backcombing to hide a bald patch – claims which were later denied.

By now I was in a strong position to consider my future with a view to severing my ties with the Met Office and pursuing a freelance career. They had provided me excellent training and career opportunities over many years but by leaving the Met Office I would also be leaving my work at the BBC, such was the inflexibility of the system.

# TIME FOR A CHANGE

My three years working on the Travel Show came to an abrupt end in 1990 when taking a day of annual leave to go to Manchester became inappropriate due to the increased workload at the BBC. Throughout the three summer series I would break my journey to visit the Manchester Weather Centre, now based in Stockport. I would gather my own data in order to present the weekend forecast for the whole of Europe as well as here at home. As a gesture of thanks I provided their Christmas raffle with a few bottles of whisky.

Their commercial arm had now decided that enough was enough and I received a letter suggesting I pay them £80 per programme for the use of their facilities. This represented over thirty per cent of my fee and was clearly a non-starter, especially as we were all employed by the same organisation and I did my own research.

Over the years the Met Office had to be more accountable. They became an MoD Executive Agency in April 1990 and started operating as a trading fund on 1st April 1996. There was also a feeling by me and a few others at the Weather Centre that they wanted more control of their television presenters. By spring 1999 stories about the BBC Weather Centre hit the press with alarming regularity as we became Big Brother and a television docusoap all rolled into one.

Two weather presenters, Richard Edgar and David Lee, put in formal complaints of bullying against the boss, Bill Giles. The outcome of seven months of written statements and deliberations by the Met Office was for the complaints to be upheld and Bill to be found guilty of being 'cruel' to colleagues. *The Sun* newspaper's front-page headline on Friday 22nd October 1999 was simply 'Cruel Bill', accompanied by a picture of a more amenable boss presenting a friendly face to viewers. Both presenters had accused Bill, amongst other things, of being an egomaniac who bullied and intimidated staff.

Publicly the lovable Ian McCaskill went a step further, having retired from the Met Office a couple of years earlier and now enjoying the freedom that retirement had bestowed upon him. Apparently he loved using his bus pass! He accused our former boss of "stalking the corridors of the BBC like a prefect at a minor public school". Working conditions for the weather presenters were "crap" and he himself had been "paid less than the doorman" for his efforts, including regular night shifts and weekend work.

The whole affair had caused an almighty storm within the Met Office and the BBC Weather Centre, but worse was to follow when Giles sought advice and decided to appeal against the decision. Amazingly just two weeks later the result was overturned. For Richard and David this was a devastating outcome from which there was to be no way back, their broadcasting careers effectively at an end. Both were summoned to Met Office HQ in early February 2000 to be told that they would not be presenting the weather for the BBC again.

The newspapers returned to this web of intrigue as the news broke with new revelations about the future of the BBC weather presenters. One senior Met Office source revealed: "We have decided that there should be a spring-clean of the team. There is a feeling at the Met and the BBC that the weather news presentation has been dogged by moaning minnies since the row over unfounded complaints by some forecasters about Bill Giles' so-called bullying style of management".

In the meantime there had been an enforced change in the management of the Weather Centre staff when Bill Giles officially retired on his sixtieth birthday, just days after being cleared of all allegations. His replacement was Helen Young, a bubbly girl with huge ambitions and plenty of confidence but limited experience. So it was no big surprise that changes were being made and by implication the average age of the team would also be drastically reduced. In my view this was an appropriate and sensible move, newspaper reports suggesting that I was against a younger team being totally unfounded, but to this day I would still favour a team with a rich blend of youth and experience.

However, to my mind morale at the Weather Centre remained low and there was clearly trouble in the ranks. The management methods within the organisation were seen as dictatorial and some of us suspected a conflict of interests. Bill Giles was also back, now employed by the BBC as a consultant within the Weather Centre. So in practice virtually nothing had changed.

My decision had been made by now, although it was a decision only known to my family.

In May, less than three months into the new regime, I was requested to meet a director from the Met Office at Waterloo station. This could have been a scene straight from one of John Le Carré's novels except I was a humble weather forecaster, not a spy coming in from the cold. Or maybe I was doing just that!

We were to meet on the stroke of midday, not under the large clock on the station concourse but outside the bar where we could meet for a drink. I had no idea what the meeting was about, it would take about one hour and then I was free to return home.

This director was ironically the same person who had been in charge at the London Weather Centre in High Holborn when I arrived from Nottingham almost fifteen years earlier. We had a good relationship and had enjoyed frank but friendly discussions at the BBC in the past. In my early years in London we had occasionally met over a pint of real ale in the local pubs for which Central London is justifiably proud.

Somehow I felt this meeting was going to be different because it had been arranged on neutral territory and there seemed to be some urgency, so much so that I was due back at the BBC that evening for yet another night shift. Of course I was on time, such was the discipline that twenty years of broadcasting instils on the presenter, but he was late – ten minutes late! He apologised, we shook hands and went inside. As usual it was my round!

On the journey to Waterloo there was a feeling in my mind that perhaps the neutrality of the venue meant I was being considered for a move to ITV Weather. Otherwise I could see no rhyme or reason for this clandestine assignation. A meeting which to my knowledge at the time had not been made aware to any other party.

As we were about to sit in one of the quiet corners of the bar I was questioned by a little old lady who was clearly a person with great powers of recognition: "Hello John, nice to see you. You are looking well. Can I have your autograph?"

"Of course you can," I replied, and we conducted a few seconds of small talk while she fumbled around in her bag for a scrap of paper. I duly did the honours and then she went to sit down in another part of the bar where I was still in her eye-line.

Clearly this was not the kind of atmosphere my director was looking for. There was a tension in the air. Maybe I was one of the "moaning minnies" and pressure was about to be put on me to consider my future.

For the next hour he wanted to see me as his subordinate and not public property to be adored by passing customers, an inevitability perhaps for a regular radio and television presenter with a strong pedigree. This was a scenario that some people in the Met Office had never quite grasped. We were household names and our screen exposure allowed us the freedom to gain notoriety and benefit from the privilege we had been granted.

What was discussed over the next hour was pure fantasy leaving me with few doubts that someone else had been party to this meeting and it was a 'stitch-up'. I was accused of giving false weather information to some of the junior weather presenters for whom I was responsible. It was suggested that I was now so introverted and presumably full of my own self-importance that I gave scant regard to the requirements of the other presenters.

What he did not know at the time, and I gave him no reason to suspect it, was that I had been considering resignation from the Met Office throughout the uncomfortable revelations of the previous twelve months.

The bullying case was just one reason why I was considering a possible career move.

I now believed the time was right and at the end of October 2000 I pinned my isobars to the wall and left the Weather Centre.

# LIFE AS A FREELANCE

On 1st November 2000 I woke with a fantastic career behind me but with little knowledge of what was in store. The feeling during the first few days was of a huge weight being lifted from my shoulders and that counted for a lot. Charlie was now eight years old, George was six. From now on my family came first.

Over many years I had made good contacts inside and outside the BBC along with sport at the highest level in this country. I was not on my own without a paddle but I had to get the boat moving in the direction I wanted it to.

On the very first day the *Mirror* newspaper was interested in doing a feature about my career and why I was resigning from what millions would have seen as a brilliant position at the BBC. So it was a pleasure to have lunch with Sue Carroll at Redcoats restaurant close to Hitchin. I talked and she listened, frequently writing notes in her journalist's notebook.I was happy with the excellent piece she submitted for publication suggesting "Kettley did not leave under a cloud" and my first fee was guaranteed.

It was important that I recouped some of the regular income I had now lost from my Met Office employment but for the first few days I needed to 'sell' myself doing interviews for local and national radio.

As luck would have it the autumn had continued to be extremely wet, probably the wettest for three hundred years, and I was asked to participate on the 'Kilroy' programme on 8th November using my expertise as a meteorologist.

An interesting and flattering opportunity had already been offered to me in local radio but I still had a vision to remain in weather broadcasting, heeding the advice given to me by directors at the BBC. Having contacted the deputy head of News and Current Affairs a meeting was hastily arranged for the following day, 9th November. The meeting was probably no longer than twenty minutes but it would open the doors to Radio 5 Live where I had already had several discussions with Philip Eden and Bob Prichard, their regular weather presenters.

In the meantime I had been approached by Asahi Beer to record a radio commercial on their behalf. So a quick trip into London on the morning of 13th November provided one of those unusual opportunities I was hoping for without my Civil Service ties. Just one hour spent in a recording studio in Soho, speaking fluent Japanese by the end of it, and I was on my way.

It was then off to meet Jim Dale, managing director of British Weather Services near High Wycombe. I had first met Jim at one of my Lord's Taverners cricket matches at Penn Street back in June. It was a chance meeting that brought us together over lunch as my kids chased his kids around the buffet table!

He founded his independent weather company in 1987 providing some competition for the Met Office. He was fully aware of my history and also my standing within media circles. Whereas he had no idea I was already contemplating resigning, it was for me another reason why I should definitely take the chance. There would be work we could do together, and independently, knowing I was the figurehead the company could benefit from.

One week later I was booked as a guest on Nicky Campbell's show at Radio 5 Live. This really was the chance to meet some of the personnel at the radio station and show Nicky and others there were other interests I could bring to the table.

By the end of my first month without a regular income I had decorated Charlie's bedroom but more importantly my groundwork at the BBC had been accomplished.

It saddened me slightly when I heard from an internal source at the BBC Weather Centre that Helen Young had released the following email to her department and my former colleagues: "John Kettley has now sadly left the Met Office and the BBC Weather Centre. If anyone calls for him you must say this. If they want him to do a TV or radio broadcast then put them in touch with the Duty Producer or myself so that someone from the BBC Weather Centre will be able to do it instead. Many thanks."

Rather than getting my solicitor to pursue the matter which was seen by some, including myself, as immoral and taking away my right to earn an income, I simply let the matter drop in the knowledge that I would be rewarded eventually.

After some negotiation I was granted a twelve months contract at BBC Radio 5 Live to work 150 weekday mornings as one of their weather presenters on a purely freelance basis. The contract would begin on Monday 5th February and coincided with a press release I had agreed with Jim Dale at British Weather Services.

On Saturday 27th January we allowed the *Mirror* sole rights to a weather story which from all known evidence at the time suggested some severe wintry conditions were imminent. We had found from experience that to sell a weather story to the press usually required slight exaggeration so the banner headline duly prepared Britain for "White Hell – Next weekend to be the coldest for 19 years".

Our evidence was the computer analysis from American and European models rather than those exclusive to the Met Office: "Siberian winds would swing across from the northeast with much of the country expecting heavy snowfalls with temperatures plummeting to minus 15C in some places. We would advise the public to keep an eye on developments next week."

I remember Steve Wright reading the Met Office forecasts on Radio 2 with no mention of an official wintry blast but he constantly referred to my newspaper article suggesting otherwise. Television forecasts from my former colleagues continued at variance with my own press release and even I was getting a little worried that I had overdone the cold blast.

Eventually most of northern Britain was hit by Arctic conditions with two hundred schools closed, scores of main roads blocked and three trains stranded in huge drifts as two feet of snow fell. In the Scottish mountains thousands enjoyed the best skiing on the slopes for thirty-five years apparently, but 11,000 homes were without power and passengers on the Lerwick to Aberdeen ferry spent forty hours at sea before reaching port.

By the end of my first week at 5 Live there had been an official comment from the Met Office saying: "This incident proves that sometimes we get it right – and sometimes we get it wrong".

Helen Young was also forced to admit on a television weather forecast that on this occasion "John Kettley had been right", despite dismissing "unofficial reports of severe conditions" prior to the event.

I must admit I was a very relieved forecaster, although for a fairly large part of the country the snow never materialised and there was plenty of rain instead. Altnaharra in the north of Scotland saw the lowest temperatures measuring minus 15C by 10th February and nearby Loch Glascarnoch never got above minus 4C even during the day on the 9th.

I had made an immediate impression in my new independent role as weather consultant and broadcaster; I could only hope that more work would be forthcoming.

Jim Dale at British Weather Services was keen to point out to the press that "the weather is the most underestimated factor in business performance". Clothes retailers planning seasonal ranges, construction companies scheduling outdoor building work and energy companies predicting consumption trends are major clients for British Weather Services. Jim went on to say: "A range of flimsy cotton dresses could be a disaster for a fashion retailer if the spring turned out to be the coldest on record."

Other projects began to massage my professional ego such as the idea of a television programme called "Britain under the Weather".

My idea was a series of six half-hour programmes looking at the history of changing weather conditions in Britain over the last two thousand years. Examples would include Julius Caesar's failed invasion in 55BC because of strong north-westerly winds and the Spanish Armada wrecked off the

British coast in 1588. Britain's worst storm in 1703 killing 8,000, the last fair on the frozen Thames in 1814 during the mini ice-age and the 'D' Day landings in Dunkerque. More recently the winter of 1962-63, the storm during the Fastnet Yacht Race in 1979, the Great Storm of 1987 and Burns Day Storm of 1990 – the list was endless.

Despite strenuous efforts no commissioning editor would give me the green light but if you don't ask you don't get.

Jim Stanton-Barber, Prime Concepts Ltd, also saw the potential in using me to produce a television game show called 'Battle of the Elements'. We first met at the Institute of Directors in London shortly after I resigned from the Met Office and clearly there were substantial rewards if a television company would buy the rights for the show. Jim worked tirelessly to sell the idea and we both visited the BBC on one occasion but the costs of such a venture and the possibility of failure meant that it has never happened.


By May 2001 my professional life as a weather presenter would again join forces with my passion for cricket as a 'once in a lifetime opportunity' landed at my feet.

Event Broadcasting, based in the Manchester area, had the contract to provide the pictures and in-house entertainment during the forthcoming Triangular NatWest Series involving England, Australia and Pakistan. Event's managing director, Alan Yardley, contacted me, presumably with the backing of the England and Wales Cricket Board, to see if I would be available as the 'pitchside' presenter. This would involve working on the grounds at eight of the ten matches in the series; the two outstanding at Lord's would not require the 'entertainment' that Event Broadcasting were offering.

A meeting was hastily arranged on 24th May with Alan Yardley and myself at the office of my sporting agent in Nottingham. It was explained to me what I would have to do at each of the eight venues and fees were agreed. It was a fantastic job and beyond my wildest dreams, although it was the kind of opportunity I was hoping for in my new 'no ties attached' role as a freelance broadcaster.

The first in the series was a day/night match on 7th June at Edgbaston between England and Pakistan. Between my duties at 5 Live I was also providing weather forecasts for 'Test Match Special' along with inserts into other television programmes for E4, Channel 4 and Channel 5.

Suddenly there were not enough hours in the day but I felt free of all the bureaucracy and, importantly, the next night-shift which would have prevented me from ever doing this kind of work before.

My preparations were complete and I walked onto the playing surface at Edgbaston, microphone in hand, about sixty minutes before play was due to start at 14.30.

It was surreal, my every move appearing on the big screen inside the ground. There I was, a presenter better known for the weather, conveying cricket information to the gathering thousands of Pakistan and England followers above the increasing sound of whistles and horns. It was a huge adrenalin rush – and I loved it.

The players were practising on the outfield close to where I made my contributions, clearly not the safest place to be as the odd cricket ball flashed along the ground close to where I explained the format of the day's entertainment and history of matches between the two sides. I interviewed several spectators and it was uplifting to be recognised by many of the gathering Pakistan followers, always vocal in their appreciation of what I had to say.

As the match got underway I could relax for a while in the Birmingham sunshine, although there was a cool breeze if you happened to be in the shade. I walked around the ground to be greeted by more recognition from officials and spectators but there was a lot more work to do between innings. More research back at Event Broadcasting's transmission lorry together with familiarisation of the quiz and other entertainments I had to oversee during the forty-five minutes break. There were more interviews to conduct in the hope that I would sound authoritative and not let anyone down.

It all went well in the end and I left the ground during the England innings feeling quite exhausted but satisfied, not to mention gaining a suntan. I travelled home listening to the commentary of the match on the radio which was unfortunately marred by some unsavoury crowd trouble.

The next two weeks saw me 'living out of a suitcase' in some smart hotels the length and breadth of England, although there was still time to get home to recharge my batteries. It was down to Cardiff and Bristol over the first weekend which were both blessed by wonderful weather. Sadly the following Thursday it was England against Australia at Old Trafford, Manchester, and the rain was back. My duties included interviewing Dickie Bird and also John Holder, one of the umpires on duty that day. Living locally these days he told me privately that his wife, who was a nurse, had met my dad when he was admitted to hospital recently – what a small world!

Thunderstorms rolled around the ground for this latest day/night encounter but there was still plenty of time for Australia to take advantage of poor batting conditions and win the match.

Two days later and the weather deteriorated further at Chester-le-Street where Australia were due to meet Pakistan. Sadly the game never started but I made many appearances on the ground informing the patient spectators what was happening, or otherwise. It was now clear to me why it was useful to have a pitchside presenter who knew what the weather was doing, or expected to do later.

It was also a quite funny that I should keep appearing to the sound of 'Don't Blame it on the Weatherman' by B*witched!

**Working as pitchside host during the 2002 Nat West Series.**

The following day was England against Pakistan at Headingley, Leeds, and the game went ahead under cloudy skies but with England suffering another defeat. The next two matches brought improved weather at Trent Bridge, Nottingham, and at the Oval in London.

My contract was complete by the summer solstice. I had travelled over 1,500 miles in rain and sunshine, met and talked to famous cricketers and spectators, had cricket balls hurled at me in the outfield and received various renditions of 'John Kettley is a Weatherman' from Chester-Le-Street to London. Above all I had a wonderful time gaining valuable presentation experience doing a job for which I was not normally known.

To finish off the week I played in Norma Major's Charity cricket match at Alconbury followed on the Saturday by the Roxy Music concert at Wembley. I went with Lynn having managed to get hold of a couple of tickets from the band when they visited 5 Live.

Life couldn't get much better than this!

The following summer I worked for a second year on the NatWest Series featuring England, India and Sri Lanka and it was a great thrill to introduce

**The scribbling on a piece of paper that Tony Blair passed to me during a programme at 5 Live Studios. Virtually illegible, it read: 'Didn't you have a rock song called after him – TB.'**

the Tenors, without the Divas this time, as part of the interval entertainment.

On a more serious note I was also required to promote the sun safety message about skin cancer. With more than 40,000 people diagnosed with skin cancer every year the series sponsor NatWest aimed to reduce that figure with a new campaign. At each of the ten matches between 27th June and 13th July the 'Sun Safety – know your boundaries' campaign was driven home to all the spectators attending at every venue. A maximum of one hundred overs of cricket in each game could mean over seven hours sitting in the sun. With that in mind 50,000 sachets of Garnier Ambre Solaire would be distributed during the series.

Of course the sun didn't always shine and like the previous year there were serious problems with the weather at Durham as rain washed out most of the England innings after the break. There was a delayed start at The Oval where England were again playing India but eventually play did get under way in warm sunshine in an overs reduced match. Yet another delayed start at Headingley for England against Sri Lanka after a morning of strong winds and driving rain. As I stood under an umbrella with Eleanor Oldroyd from 'Test Match Special' it looked very unlikely that any play would be possible but my latest radar information from back in the pavilion suggested all would be well eventually – even the result as England won in the final over.

An abiding memory of my last match at Bristol was appearing on the pitch as usual to introduce the Tenors performing Nessun Dorma only for them to break out with a personal rendition of 'Happy Birthday To You'. Yes it was my final NatWest game and also my birthday and I was treated to magnificent tribute on the big screen and all around the ground – truly moving!

One of the strangest requests I have received in recent years was from Alistair Campbell who called me one morning from Downing Street. We had met on many occasions in the past going back to his days as a columnist at the *Daily Mirror* when he came to 'BBC Breakfast' to review the morning newspapers. Together we spent time over a coffee talking about football, more specifically Burnley FC.

On the phone he referred to the Afghanistan war and how the Government was concerned about how weather conditions could play a part in their operations. Of course I agreed that the weather was significant, especially so in the winter months and particularly in the mountainous areas where some of the troops found themselves.

Alistair was keen to know if I could provide weather forecasts to aid operations, probably believing that I was still in the Met Office which was still an integral part of the Ministry of Defence.

I explained that the Met did have a selected group of personnel to call on for wartime situations and, although flattered to be considered, this was really a situation beyond my normal 'comfort zone'.

My expertise was still in broadcasting and it was a pleasure to meet people on a regular basis when they visited the 5 Live studios. After years of becoming more isolated at the Weather Centre I was now enjoying friendly banter again with national celebrities and sporting icons. Actors and politicians often appeared on Nicky Campbell's morning programme and it was during the General Election campaign in early June 2001 that Alistair Campbell and Tony Blair appeared. The only available seat in the studio was next to the Prime Minister so there I sat. We acknowledged each other while a caller asked a question. During a pause in proceedings with the microphone still 'live' he scribbled on a piece of paper and passed it to me. Virtually illegible, but full marks for his musical knowledge, the scrap of paper was asking if I was the weatherman who had a song written about him.

I nodded feeling quite proud but was inwardly pleased that he had other people to write his letters for him, otherwise they would be virtually unreadable!

Consultancy for the travel industry, weather sensitive businesses and sporting venues were other main interests.

It was early in 2002 that my first breakthrough came into horse racing, not because I thought myself light enough to become a jockey but as a weather consultant for some of the major racecourses in Britain.

The Cheltenham Festival generally falls in the racing calendar around mid-March, although in 2001 the meeting had to be abandoned due to the devastating foot and mouth epidemic. Their clerk of the course, Simon Claisse, made contact shortly before the Festival informing me that the last five months had been very dry with only nine inches of rain falling on the course.

Desperate times for National Hunt Racing's premier event required desperate measures. They needed to pour lots of water on the course to keep the ground 'safe' for jump racing but I was asked for my opinions (as a freebie this first time) about how the weather was looking in the coming days and during the three-day meeting itself. I rightly indicated that some wetter weather was likely to help their cause but it would also turn much colder as the winds switched into the north by the third day.

Heartened by my accuracy – the temperature began at 13C on the first afternoon but was down at 7C by the third day – I was granted a contract for the following season, 2002-03. Since then my relationship with Edward Gillespie, Managing Director of Cheltenham Racecourse, and Simon Claisse has gone from strength to strength. They have learned from me and I have certainly improved my understanding of how a bespoke weather service for an area little more than fifty acres in extent is so much more challenging than a television weather presentation could ever be.

Long before my involvement with Cheltenham Racecourse the Festival had to be re-scheduled in 1978 due to a sudden heavy fall of snow. On the Wednesday evening officials had been assured that all would be well for racing the following day but the clerk of the course woke to an eerie silence in the early hours of Gold Cup morning, March 16th. To his astonishment four inches of snow had fallen in a short time and the great day was ruined.

Julian Wilson was working for BBC Television by now and he remembers the abandonment being a complete surprise. He recalls: "On the Wednesday a dramatic Champion Hurdle was won by the heroic Monksfield, from his perennial rival Sea Pigeon. The weather was cold and wet, but very raceable; the going officially 'Good', and nothing prepared racegoers – or broadcasters – for what was to come. Scripts had been written; build-up items filmed and edited; interviews planned. All was arranged for a BBC-TV transmission from 2.15 to 4.35.The broadcasters slept soundly…only to awake at 7.00.am to find the surrounding Cotswold hills covered in several inches of snow!

"When the race was eventually run four weeks later, the field had changed shape significantly after the intervening Grand National, and Irish Grand National, won by the twelve year-old Brown Lad. The favourite was another twelve year-old Fort Devon, but, in mid-April, the going (now 'Good to Firm') had turned against him, and the race was won by the doubtful stayer Midnight Court, whose season's record was now seven wins from seven starts. Brown Lad was second.

"There was dramatic irony in the result. The winner was ridden by John Francome, who, on the eve of the abandoned Gold Cup four weeks earlier, had been interviewed by the chief of racecourse security over allegations that he had been 'stopping' horses in collaboration with the flamboyant

The Cheltenham Festival in full gallop in March 2007. It has been a great privilege to have provided the event with a bespoke weather service since 2002. (EMPICS Sport)

Scots bookmaker John Banks. The racing press had a field day with the story, and the following week, on BBC's 'Sportsnight', I ran an exclusive item, showing a number of races in which Francome's riding was looked at.

"It was with trepidation that I approached Francome for an interview after his Gold Cup win – but he bore no malice. In the event Francome faced the Jockey Club Disciplinary Committee on April 30th; was found guilty of breaching Rule 220 (ii) (providing confidential information); was fined £750, and suspended for six weeks. All agree that Midnight Court would not have won the Gold Cup on soft ground on the original date."

There have been other close calls since then, especially in 1987 and 1989. On the first occasion the Gold Cup was delayed as snowstorms swept over the course. After the previous race, the Foxhunters' Chase, the snow intensified, and after the runners had reached the post they were sent back to the paddock and the race was put back from 3.30 to 4.50 – "not a popular decision with BBC-TV", according to Julian.

**Pre-race advice to the Director of Racing at the Cheltenham Festival.**

Two years later Desert Orchid triumphed on muddy ground following morning rain and snow. I had been on television duties that morning for BBC Breakfast, reporting live on several occasions from the grandstands overlooking the winning post. The cold went deeper into my weakening body due to a notable absence of hot tea and coffee, but immediately the programme closed at 9am I tucked into a hearty cooked breakfast in the media canteen.

No meal could ever have been more appreciated but my worry was that the meeting could eventually be abandoned as the rain turned to wet snow and the temperature continued to fall. The surrounding Cotswolds were slowly turning white as the snow settled above this most natural amphitheatre. In the end all was well but for Desert Orchid it was a punishing victory on ground he must have hated.

The weather forecast ahead of the Cheltenham Festival is always very high profile within the racing industry and the Racing Post keeps its readership in touch with the daily forecasts. As they have a contract for weather information with the Met office there has often been something of a stand off between their forecasts and the official predictions I provide through Simon Claisse. My expertise is well respected by all at Cheltenham so they pay scant regard to what they quote, although it does lead to explanatory conversations between the two of us.

In 2009 the difference of opinion was more extreme than usual as the Festival approached because the *Racing Post* continually quoted rainfall expectations far in excess of what I anticipated. On Saturday my original suggestion that up to 10mm of rain could hit the racecourse between then and the first race on Tuesday was challenged by a 'Met Office spokesman', who indicated that my predictions "could err on the conservative side". Indeed they were sticking to their view that 15mm to 20mm was expected on Monday night alone as a period of heavy rain was expected to sweep through the area.

Speaking to the *Racing Post* on Monday afternoon my estimate was of just 5mm to 6mm between 2am and 6am on Tuesday with the Met Office still adamant that 15mm to 20mm was expected. The meteorological 'bun fight' was widely acknowledged amongst the racing press but I was undeterred, as indeed were the Cheltenham officials, who repeatedly confirmed that I had the greater experience after many years working on their behalf.

By 7am on Tuesday the Racing Post confirmed: "Cheltenham is officially good to soft, soft in places, for the first day of the Festival, after just two millimetres of rain fell overnight. On Monday the Met Office had predicted that ten to twenty millmetres of rain could fall overnight, while Cheltenham, who use private forecaster John Kettley, had anticipated only five to six millimetres of overnight rain."

Of course I was proud of my efforts and fully understood why other forecasts had been wide of the mark but once again my greater understanding of the microclimate of Prestbury Park proved invaluable and must have been more useful to punters and trainers as a result.

Occasionally I have to try and help in more bizarre circumstances. When the sun is low in the sky during the race meetings at Cheltenham between November and January there are safety issues as the jockeys and horses have trouble negotiating the fences in the home straight.

A review was carried out in 2005 with the Sports Turf Research Institute in which the most dangerous period was estimated to fall between 14.40 and 15.50. Ideally Simon Claisse would prefer cloudy afternoons after a sunny morning to maximise crowd attendances but adjusting the times of some of the races has proved necessary in recent years. A second option would be to erect a portable screen to block out the sun but that could prove too costly.

As part of the select band of racecourses known as the RHT, now Jockey

Club Racecourses, it wasn't long before I was speaking to other administrators within the same group of fourteen.

My first contact at flat racing's headquarters in Newmarket was with their Director of Racing, Michael Prosser, in early March 2003. An agreement was soon forthcoming which has been in operation ever since to the great benefit of both parties.

Expected timing and amounts of rain, severity of frosts in the winter months and strong winds are all part of the rich tapestry of work and agreements I have with individual racecourses. For flat racing the major issues during the season tend to be based on the watering schedules of the track in order to keep racing as safe as possible with no jarring for the horses, thereby minimising injury. Evapotranspiration governs how much moisture is taken out of the ground and depends greatly on time of the year and weather conditions. It is a science understood by the Directors of Racing and it is up to them to make best use of the forecasts I give them throughout the season.

They also need to utilise the forecasts for the management of the turf, including application of fertilisers at the appropriate times when the weather is suitable.

To be somehow part of top class sport in this country is hugely flattering and satisfying when all goes well. Over the years it has been a great privilege to have expanded my portfolio of sporting venues, either through word of mouth or direct contact.

I feel very proud to have been the major weather consultant for several years now at Aintree for the Grand National meeting, Haydock Park and Ascot for both flat and jump racing throughout the year, and the England and Wales Cricket Board for Test Matches in this country.

Keith Kent is head groundsman at Twickenham – home of the Rugby Football Union. After fifteen years at Old Trafford, where he saw Manchester United become the top team in the Premier League, he accepted the "job he couldn't refuse" in September 2002. He wrote an article on the Pitchcare website shortly afterwards appropriately entitled 'John Kettley Is a Weatherman'.

In the article he reflected on our early work together when he had just agreed a contract with Sports and Stadia Services (SSS). There is no undersoil heating at Twickenham but SSS can provide additional covers and heating to protect the pitch when severe frosts are expected.

Prior to the Australia game it was not so much the frost but the fog that was concentrating his mind that Saturday morning in November. He could not see across the pitch and was becoming increasingly agitated that the game could be called off if conditions didn't improve. Keith called me and I assured him the sun would be shining brightly before kick-off – and it was!

**Forecasting barbecue weather can be dangerous – as the Met Office discovered in 2009 – but this did not deter me from launching the John Kettley BBQ Index.**

The following weekend was England against South Africa and the rain was so bad locally that Ascot races were abandoned. Again Keith was on the phone and I was very hopeful that morning showers would clear then nothing more until after the match.

The rest of the day was as forecast and I had just got lucky. For Keith I could do no wrong and we have worked together ever since. I have also interviewed him on my local radio stations when the weather has been a point of discussion and he continues to ring my praises. Good man Keith!

My association with Sports and Stadia Services goes back to a Lord's Taverners cricket match at Bishops Stortford in August 2002 where I first met Nigel Felton, former Somerset and Northamptonshire opening batsman. He was now in business with Allan Lamb, a team-mate who also played seventy-nine Test Matches for England. Their business was now to protect pitches against the worst of the British weather, particularly severe frosts and heavy rain. We chatted together at great length whilst fielding and it was clear that I could provide the kind of weather information he needed on a regular basis and at the same time add authority using my name.

The working relationship we cultivated has continued to thrive and expand over the years so that I now provide regular information to the Football Association as well as all the fourteen courses under the stewardship of Jockey Club Racecourses during the winter months.

A contract with Birmingham City Football Club was also secured as a direct result of working with Nigel at SSS. In January 2007 this relationship became public knowledge when it reached the sports pages following a week of terrible weather in the West Midlands.

The pitch at St Andrew's was originally scheduled to be re-laid during the week commencing 18th December 2006 but in the light of my forecasts of freezing conditions a delay was agreed until Sunday 7th January. Although I was now suggesting rain every day it was agreed by all concerned, including the contractors, that anticipated rainfall amounts would not cause interference to the operation.

Regular updates of worsening weather conditions were sent to the Club through the week and with the passage of time the pitch became saturated and increasingly difficult to work upon. The result was that the next Championship match against Leeds United on Saturday 13th January had to be postponed as it was not ready for use.

The Football League later charged Birmingham City FC with misconduct because they could not fulfil the fixture and a hearing was convened at St Andrew's Stadium on Friday 20th April 2007. I was invited to give evidence along with Steve Bruce, team manager, Karren Brady, managing director, and other officials from Birmingham City FC together with representatives of the companies involved in the pitch re-lay. This was a new experience for me but I felt the club had a strong case due to rapidly changing weather developments which I emphasised during the questioning.

Following the hearing the Football League Disciplinary Commission cleared the Club of misconduct and issued this statement: "The postponement of the game was caused by circumstances beyond the control of Birmingham City, namely freak weather conditions. Those conditions could not reasonably have been foreseen or anticipated before the pitch was taken up and occurred before the relaying could be completed."

Karren Brady personally thanked me for my evidence and we continue to enjoy good relations.

Away from the sporting environment it is always flattering to be invited to represent companies at the launch of a new product or to endorse them. Many have come my way in recent years since those early days of Asahi Beer and they can range from the bizarre to the scientifically challenging where plenty of expertise needs to be incorporated.

The John Kettley BBQ Index was launched for the summer of 2003 by Dalepak / Ross Burgers in collaboration with British Weather Services. Sadly the weather during July and August did nothing to aid the success of the project but there was no lack of commitment on all sides.

The Index was based on seven weather elements – temperature, wind speed, wind direction, humidity, sunshine, available daylight and rainfall. The weighting of each element in the range 1 to 5 meant that a maximum

**Television has continued to play a major role in my freelance career. In September 2007 I took part in Ant and Dec's 'Takeaway' on ITV.**

score of 35 was possible. We believed that the perfect BBQ conditions required:

*Dry conditions with a temperature between 22C to 25C.*
*Sunny skies, wind speed around 5mph from the south.*
*Low humidity and 12 hours or more of daylight.*

The success of such a project requires comprehensive knowledge of the weather together with a well-known name and a long session in the radio studio to spread the word across the country. Very tiring, but always rewarding, such product endorsements have become a welcome departure from day to day forecasting through Media companies such as Finn based in Leeds where the managing director, Richard Rawlins, has been very supportive.

One of the more bizarre was a photo-shoot on Derby railway station advising on sun protection in summer 2004. It was part of an advertising campaign by Midland Mainline to encourage people to take advantage of the hot summer weather and the great deals they were offering on tickets to London.

So a photo-shoot in Derby together with a radio commercial to record at the studios of Chiltern FM in Luton made for another satisfying venture but

**The draining experience of appearing on 'The Weakest Link' in the summer of 2008.**

even that was surpassed by the offer of a television trail later that week.

I was approached by MTV, one of Britain's premier music subscription channels, to record a set of trails at their studio in Camden. They asked me to dress as I would have done in my former weather presenting days in a smart suit, weather charts behind me and record several different versions to encourage the kids to watch MTV that summer because the weather would not always be good outside.

Quirky and challenging it certainly was, but back at the BBC many people told me they had seen my trails while they were working-out in the gym with MTV playing in the background. It was almost as though I had broken back into television weather presentation, as long as the sound was turned down!

I owe a tremendous amount to the many people who have provided me with working opportunities since I took the brave step of resigning from what millions would have seen as a safe position. I have never regretted my decision for one minute.

Radio and television have continued to play a major role in my freelance career with the chance to work again with old friends or media icons. Over the years I have sparred with Anne Robinson on three different programmes.

During her time on 'Watchdog' I was once asked to present weather details for the Mediterranean, on behalf of the consumer, in connection with an offer from Virgin Travel.

Then in March 2005 I was a contestant on 'Test the Nation', a live Saturday evening quiz programme on BBC1 which Anne co-hosted with my old pal, Philip Schofield.

Finally, in the summer of 2008 I got the call to appear on 'Weakest Link' which was probably the most draining of the lot. You feel enormous pressure to do well once you get past the fear and embarrassment of exiting in the first round. Thankfully I managed to stutter my way through to round six which left the way open for Dermot Murnaghan to become the worthy winner. One programme requires almost three hours of filming, including numerous breaks for the researchers to brief Anne on the contestants and the performances in the last round. I was grateful for the easy ride she gave me, as she did on 'Test the Nation', so for me she is a pussycat!

Since leaving the southeast for the slower pace of life in the Lincoln area my purpose-built home radio studio has provided greater flexibility to service the needs of local radio stations, BBC 5 Live, BBC Asian Network, Radio Wimbledon, Talksport and Radio 2 on many occasions – and long may that continue as my enthusiasm for broadcasting remains as great as it ever was.

In summer 2009 Radio Wimbledon used me for the seventh consecutive year and on this occasion I managed a day at the Championships which was also a memorable opportunity. I had presented television weather forecasts on several times in the past but needed to leave before play started in order to complete my shift at the BBC Weather Centre.

In 2002 I also appeared on Centre Court with Tim Henman – how many people can say they managed that! Actually we were not allowed on the grass as it was less that one week before the Championships were due to start so a photo shoot on behalf of Slazenger was conducted in the first row of seats. Tim was a gentleman and it was a great pleasure to have met him, albeit for just under an hour.

What I didn't expect in my new freelance career was the chance to write in a national newspaper but that is exactly what I have been doing since February 2005.

*The Mail On Sunday* approached a small group of people, including Michael Fish and myself, to trial a 400-word weekly column for their Features section. Initially Michael was given the job but after a six week trial the batten was handed to me and I have been delighted to write for the newspaper ever since. Always topical and occasionally confrontational, this small step into journalism has given me the confidence to continue writing.

Who knows how many more surprises lie ahead in my ever-changing world as a broadcaster and weather consultant.

# VIEWS ON
# CLIMATE CHANGE

There was a certain irony in the fact that in 1974, just ahead of two hot summers, the BBC produced a documentary programme reflecting the threat of 'global cooling' targeted by some scientists who were already anticipating the next ice age.

Nigel Calder, father of the travel writer Simon Calder, was the presenter who had previously held a position as editor of *New Scientist*. This corresponded to the time I was based at the Meteorological Research Flight in Farnborough where their modified Hercules aircraft was currently undergoing research in Senegal as part of the Global Atmospheric Research Programme. This project was being developed to gain a better understanding of climate variability, the emphasis in this case being in the tropics.

In those distant times politicians were generally in the sceptic camp, understandably neither convinced that the Earth was warming-up nor cooling down. 1977 saw the First World Climate Conference, but at an informal meeting convened by the World Meteorological Organisation during their preparations, one representative was quoted as saying: "Politicians will never take any action about climate change until meteorologists can tell them what will happen rather than what might happen."

Fortunately by 1979 we had a Prime Minister with a scientific background who was in touch with influential scientists in this country, including those in senior positions at the Met Office.

During the following decade more evidence was gathered on climate fluctuations at home and around the world, not only from season to season but also decade to decade. Margaret Thatcher was on a mission and along with other leading heads of government she attended the Second World Conference on Climate Change in 1990.

During the 1980s our increasing knowledge of the atmosphere meant there was now a greater consensus that the world was getting warmer rather than cooling down and for Margaret Thatcher this was a 'double whammy'. After the coal miners' strike in 1984 she could continue to run-down the coal industry in this country armed with the latest scientific information that the burning of coal was greatly adding to carbon dioxide emissions and enhancing global warming. Remember that carbon dioxide, along with methane and water vapour, is a 'greenhouse' gas which allows direct sunlight to shine through but retains some of the heat reflected back from the surface of the Earth.

The subject of climate change, with an emphasis on rising temperatures, has become a 'sexy' subject over the past twenty years dating back to the original talks in Kyoto between many of the leading industrial nations. By that time many people were increasingly convinced that the burning of fossil fuels and clearance of forests had increased the amount of carbon dioxide in the atmosphere by a third over the past 150 years.

It should always be remembered that climate and the weather never stands still and it has always gone through periods of warming and cooling for different reasons. If greenhouse gases did not exist the world would be at least 30C colder so could not support life as it does today. There is a certain natural variability about our climate, a sinusoidal pattern of temperature through the decades and centuries, with spikes and troughs along the way such as the harsh winter of 1963 and the hot summer of 1976.

Ice Ages occur irrespective of how many cars are on the roads or light bulbs we use – they result from a natural change in the orbit of the Earth around the sun, so altering the amount of solar energy striking different parts of the world. Thankfully the last major Ice Age occurred just over 10,000 years ago and they are reckoned to have a cycle of about 100,000 years!

Sea levels rose between 5000BC and 3000BC as the ice sheets melted and Europe's temperatures were said to be 2C to 3C higher than they are today. The arrival of the Romans in the first century AD was during a period of colder weather but Britain again warmed considerably during the next four hundred years. During that period vineyards became widespread across Britain as far north as Yorkshire, indicative of the higher temperatures which made Britain self-sufficient in wine.

After a colder spell further medieval warm periods meant that summer temperatures were again possibly higher than they are today and from the 12th to 14th centuries even Greenland became mostly green. The so-called 'Little Ice Age', which roughly spanned the period from 1500 to 1825, saw London street fairs become a regular event on the frozen Thames between 1607 and 1814.

Slow warming began again, maybe as early as 1730, with the imminent industrial revolution inevitably causing some form of climate change as copious amounts of smoke billowed into the atmosphere from industrial chimneys and rows of terraced houses.

During the early part of the 20th century, and again around the time of the Second World War, Britain saw warmer summers on a regular basis before it turned generally cooler again through the 1950s to the early 1970s. The harsh wintry weather in the 1980s occurred in spite of the realisation that the world was seeing the start of what we now are told is unprecedented

'Global Warming'.

We should be aware that what we experience these days at home and around the world is nothing new from the point of view of dramatic weather events such as hurricanes and tornadoes. It is more significant that climatologists, with all their powerful computer models, are now saying that we have not seen the like of it before and the various scenarios for the next century are potentially more destructive. I have to believe that myself but the problem lies in how the information is released through the media.

1998 followed by 2002 remain the warmest years globally on the current data sets which go back less than 160 years. Of course we hear of severe droughts and forest fires in Australia and California, along with the record-breaking heatwaves in Britain during August 2003 and July 2006. Individual spells of extreme weather in different areas are mutually exclusive to the overall pattern of world climate change.

As I said earlier it has become a frequent topic of conversation throughout the media, almost to the point that it can be ridiculed. Sometimes I feel a football manager has lost his job because of 'climate change' because there is really no other reason why he faced the sack.

Every few months the latest report from one climatologist or another, or even the highly respected Intergovernmental Panel on Climate Change, maintains that the situation is now largely human-induced. Even allowing for the fact that the greenhouse gases have always been there, should we really be made to feel so guilty when we haven't yet switched to low-energy light bulbs or electric cars.After all, no one knows for certain just how bad things will be in one hundred years' time.

Worse still are the projections which vary enormously from 2C to 5C! Do they mean that every summer will continue to get hotter as each decade passes and snow will be confined to the history books. Probably not!

All right I am playing devil's advocate to some extent, but stories make the newspapers on a regular basis which can only harm the scientific community because of the way the subject is being sold to the public:

"Strawberries in December. Frogspawn in February.

Not 2007 but 1868 – and no one blamed global warming then."

This was a headline in the *Daily Mail* on Friday 2nd March 2007 referring to the mildest winter in Central England since 1659. The early signs of spring at Haughton Hall in Shropshire were the snowdrops on 7th January and the crocus in flower on 31st January.

As a weather forecaster and presenter, the words I can understand from the climate scientists are that the "glaciers are melting and the ice sheets of Antarctica and Greenland are slowly losing mass and contributing to sea level rise. Greenhouse gas emissions continue to rise globally at a rate of 1% per year".

I am much happier to learn of those facts than for the latest severe weather event to be blamed on global warming. The Boscastle flooding in 2004, the Birmingham tornado in 2005 and the severe summer floods over England in 2007 followed by another poor summer in 2008, were events which could have happened one hundred years ago!

It is not acceptable to hear of the "highest temperature on record" only to realise that it is actually the highest temperature since it last occurred back in 1895! I am committed to global warming but I simply have no idea to what level.

More significantly we should all try to do something about it if we can although when China continues to build coal-fired power stations at the rate it does you wonder why we should bother. Well, we must bother for our future generations!

On 15th April 2002 I chaired a meeting for the Buckinghamshire Federation of Women's Institutes in High Wycombe entitled 'Water, Weather and Waste'. Speakers from the Environment Agency, the British Antarctic Survey and an Environmental Health Officer contributed to the meeting which highlighted uncertainties about future climate prediction but at the same time stressed the need to be aware and act accordingly.

Waste and green energy are fundamental to the future of our planet. There will always be an element of 'not in my backyard' but I am a great advocate of alternative types of energy production such as wind turbines, solar panels and biomass.

Going back to my university days at the time of 'global cooling', significant research was then being undertaken at Edinburgh University with a view to harnessing energy from wave power. It was so revolutionary that BBC 'Tomorrow's World' filmed the basis of their idea in the mechanical engineering department of Lanchester Polytechnic, Coventry, who were now associates of the project. It made sense to make use of the hundreds of miles of coastline around the British Isles to generate electricity from wave power, although costs were probably a major stumbling block.

In the intervening years wind turbines have become increasingly utilised along many coasts and also in the hilly inland regions where they are seen as less intrusive. People who live nearby say they are quite noisy but I have yet to prove that. However, aesthetically I believe the white turbines beautifully blend into the environment and I see them as hugely beneficial.

We have very few days in this country when the wind doesn't blow but if the sun is shining we can make more use of solar panels, either domestically or in offices and factories. Of course there is no guarantee we will receive more hours of sunshine in the next century but as technology improves the photo-voltaic cells will act more efficiently in bright weather and not necessarily when the sun is beating down.

Costs for installing domestic wind turbines and solar panels remain prohibitive for many households but personally I have another issue. We

**Participating in the Ecobuild forum 'Changing the World without costing the Earth', February 2008.**

live in a barn conversion with grade two listed status so it is possible that permission to install such devices, even when we are helping to save the planet, could be refused. My county council in Lincolnshire is commited to save millions of pounds in the fight against climate change. In a move for a more sustainable future they submitted a planning application in 2009 for a new waste treatment facility which provides the technology to turn waste into energy. This is the household waste left over after re-cycling instead of sending it to landfill. Good news all round because a landfill site generates methane, one of the major greenhouse gases, which could be eliminated at a stroke. The waste will be burnt under strictly controlled conditions to generate energy in the form of electricity and steam.

Of course we still have the problem of cows in the countryside creating copious amounts of unwanted gas!

In February 2008 I was invited to participate in a discussion at the Ecobuild exhibition in Earls Court, London. Based on the 'Question Time' format the snappy title was 'Changing the World without costing the Earth', chaired by James Naughtie from Radio 4. Although not strictly my area of expertise I was pleased to contribute along with Jim Dale – British Weather Services, David Strahan – documentary film maker, and Colin Butfield – Head of Campaigns, World Wildlife Fund.

For a whole hour we thrashed out various aspects of climate change, renewable energy and the way forward. My standpoint was, as I have described earlier, eighty per cent committed to climate change but the remaining twenty per cent of me not exactly sure whether it will be the worst case or best case scenario postulated by those experts far more knowledgeable than I.

There will always be a smaller group of scientists and academics who refuse to accept that we are in danger of seeing the problem escalate through the century. Personally I respect their views and they should be listened to because how do we really know what will happen.

Some are keen to use sunspot activity as a guide to how the climate might change and they can highlight events during the 'little ice age'. They will say that some of our severest winters in 1683/84, 1794/95 and 1813/1814 corresponded to periods of low sunspot activity, which is also the case at the present time. The second coldest winter on record in 1740 followed a warm decade during the 1730s and some scientists liken our recent period of warming to what happened then.

They are inclined to the opinion that colder winters are on the way back in the next ten years because there are no sunspots on the sun right now. That is in contrast to the high level of sunspot activity throughout most of the past twenty years which have largely been dominated by mild winters.

Severe winters were common from the 16th to 19th centuries but that did not preclude some mild winters and hot summer sunshine. We are back to the peaks and troughs within the general climatic trend. In fact 'climate change' often brings big swings from one extreme to the other either with temperature or rainfall.

Another group of scientists postulate that our current warming will actually turn the weather colder eventually because the Gulf Stream will fail. We know that this conveyor of warm water stretching out of the Gulf of Mexico and across the Atlantic helps to keep us milder or more temperate than we would be otherwise.

If the Arctic ice continues to melt it will introduce more fresh water into the north Atlantic and change the salinity of the ocean, thereby altering the strength and direction of the Gulf Stream as the density of the water changes. They suggest a worst-case scenario of the Gulf Stream breaking down in mid-Atlantic which could lead to a drop of perhaps 5C in the UK temperatures over two decades.

I only hope that the large number of reports being issued by scientists is not simply to justify the vast amounts of money thrown in their direction for their research.

All I ask is for reasoned and sensible conclusions and not for them to believe what they want to believe for political reasons.

There is no doubt that whatever happens from day to day and year to year, the weather will continue to fascinate and make headline news.

# EPILOGUE

I began with a look at the unusual winter of 2008-09 following a succession of mostly mild winters since the turn of the century. Was it a 'blip' or can we expect more of the same, despite the forecast of continuing global warming in the coming decades?

As a weather forecaster and broadcaster I have a vested interest in what happens but I do not have the answer. The experts tell us that a warmer world is generally a wetter world, though not everywhere at the same time.

In Britain the thinking would be that autumn and winter would become milder and wetter. Weather systems would become more vigorous because of the extra energy available to fuel the storms. North-western parts would have more frequent gales and heavy rain but south-eastern areas would more likely suffer drought conditions due to more frequent hot weather in the summer, punctuated by severe thunderstorms. There will be no sudden change and every switch in the wind direction will continue to bring its own type of weather.

As millions of us take holidays in the Mediterranean every year we expect the weather to be the same when we get home, and it rarely is. For example, 'Flaming' June is hardly a true reflection of what we usually see in the British Isles – just ask regular visitors to Glastonbury.

A few excellent summers over my lifetime have been 'the exception to the rule'. The dreadful summers of 2007 and 2008 are hopefully not a sign of things to come but our summers are largely governed by the Azores high pressure which we need to move north to bring a settled, warm spell. When the jet stream high in the atmosphere sinks further south we are under threat from heavy rain bands and possible flooding.

From the point of view of weather consultancy these are interesting times as changing weather patterns will inevitably impact on different walks of life and businesses in varying ways. Climate change could alter the way we live and where we live and work. It could bring new health problems either from higher temperatures locally or the spread of diseases from other parts of the world.

At British Weather Services we realise that businesses could be financially more secure whatever the weather. Now businesses can insure against 'unexpected or unusual' weather in the years to come and protect their profits.

No matter how this and future generations see the weather impact on their day to day businesses and leisure activities we will all have to live with it – but we can also gain from it.

# POSTSCRIPT

We do our best at school but develop and mature in our own time.

Some take the fast track where academia suits perfectly and no task is too onerous, all exams fall within their compass and the sky's the limit – they thrive in the outside lane of a motorway!

Millions take it steady, hogging the middle lane but feeling comfortable in their environment. Not pushing themselves too hard but getting by and performing conscientiously without going that extra mile to achieve what may be at the limit of their ability.

Many take the safe and easy option, cruising through school at a gentle pace knowing that moderate qualifications may be sufficient in the end, although never truly performing to their full potential. Maybe they can't see the point of studying and simply want to enjoy life outside school to the full, or they are just not maturing sufficiently to do themselves justice.

A few, the small minority, are happy on the hard shoulder where they care little what is happening around them but just behave anti-socially and make as little effort as possible.

Whichever route is taken we should make the most of our lives and take the breaks when they come along.  I was happy travelling in the middle lane and I believe made the most of the breaks which came my way.

Former BBC Weatherman Jack Scott was my mentor to whom I will always be eternally grateful: "Keep a diary – you never know when you might need it."

My dad who helped make everything possible and gave me the encouragement to develop the way I did.
RIP

*John Kettley*
*September 2009*

Freelance opportunities
continue to be many and
varied. In June 2009 I was
doing a weather-related
promotion for EasyJet.

# ACKNOWLEDGEMENTS

Books and articles:

Paul Hudson & Ian McCaskill, 'Frozen In Time' (Great Northern, 2006).

Ray Simpson, 'Clarets Chronicles' (Burnley Football Club 2007).

Malcolm & Freda Heywood, 'Cloth Caps & Cricket Crazy' (Upper Calder Valley Publications 2004).

Joyce Kay & Wray Vamplew, 'Weatherbeaten' (Mainstream Publishing 2002).

Robin Stirling, 'The Weather of Britain' (Dlm 1997).

Roger Birch, 'The Todmorden Album' (The Woodlands Press 1983, 1987, 2007).

Alfred Glenn, 'Weather Patterns of East Anglia' (Terence Dalton Ltd, 1987).

Dick Francis, 'Second Wind' (Michael Joseph 1999).

Stuart Whatley, 'Arctic Buxton – A Century of Winters'.

O M Ashford, 'Change of Political Climate' (Weather, January 2002).

Mark Davison & Ian Currie, 'Surrey in the Hurricane' (Froglets 1988).

Geoffrey Moorhouse. 'The Long Winter 1962-63' (Guardian).


Julian Wilson

Jim Dale

THUNDER IN THE MOUNTAINS
The Men Who Built Ribblehead
By W.R. Mitchell

The railway settlement on and around Batty Green, at the headwaters of the Ribble, lived, throve and died in less than ten years. This beautifully produced book focuses on the lives of the workforce, their wives and children, who experienced earthquake, flood and an outbreak of smallpox. Read about their austere lives, about saints and sinners. A railway missionary preached the Gospel and organised "penny readings". Policemen and excise men snooped around the huts at night, detecting illicit drinking of beer that retailed at sixpence a quart.

WAINWRIGHT – Milltown to Mountain
by W.R. Mitchell

This ground-breaking, richly anecdotal and personal book about Wainwright also recalls his young days in the Lancashire milltown of Blackburn and his fascination – as a lone walker – for wild places in Lancashire, along the Pennines, which have been described as 'the backbone of England', and in the north-west extremities of Scotland. He devised the popular Coast to Coast Walk, from the Irish Sea at St Bees to the North Sea at Robin Hood's Bay.

FIRE AND ASHES
How Yorkshire's finest took on the Australians
Introduction by Geoffrey Boycott

There are eighteen Yorkshiremen still alive who, whilst playing for Yorkshire, also participated in the Ashes Tests. In this book they recall favourite memories and stirring moments in their own words. Collectively they reveal how Yorkshire's finest took on the Australians in one of international cricket's most famous rivalries. The memories go back to the early 1950s with Brian Close and Bob Appleyard, and range over the next half century up to the 2005 Ashes that included Michael Vaughan and Matthew Hoggard.

ENGLISH JOURNEY
By J.B. Priestley
Special 75th Anniversary Edition

In 1934, JB Priestley published an account of his journey through England from Southampton to the Black Country, to the North East and Newcastle, to Norwich and home. In capturing and describing an English landscape and people hitherto unseen in literature of its kind, he influenced the thinking and attitudes of an entire generation and helped formulate a public consensus for change that led to the formation of the welfare state.

Prophetic, profound, humorous and as relevant today as it was 75 years ago, English Journey expresses Priestley's deep love of his native country and teaches us much about the human condition and the nature of Englishness.